LINEAR PROGRAMMING
Basic Theory and Applications

McGraw-Hill Series in Quantitative Methods for Management

Consulting Editor

Martin K. Starr, *Columbia University*

McKenna: Quantitative Methods for Public Decision Making
Swanson: Linear Programming: Basic Theory and Applications

LINEAR PROGRAMMING
Basic Theory and Applications

Leonard W. Swanson
Graduate School of Management
Northwestern University

McGraw-Hill Book Company
New York St. Louis San Francisco Auckland Bogotá Düsseldorf
Johannesburg London Madrid Mexico Montreal New Delhi
Panama Paris São Paulo Singapore Sydney Tokyo Toronto

To Winifred and Sue

This book was set in Times Roman.
The editors were Charles E. Stewart and Madelaine Eichberg;
the cover was designed by Carla Bauer;
the production supervisor was Gayle Angelson.
The drawings were done by J & R Services, Inc.
Kingsport Press, Inc., was printer and binder.

LINEAR PROGRAMMING

Basic Theory and Applications

234567890 KPKP 83210

Library of Congress Cataloging in Publication Data

Swanson, Leonard W
Linear programming.

(McGraw-Hill series in quantitative methods for management)
Includes index.
1. Linear programming. I. Title.
T57.74.S93 519.7′2 79–10092
ISBN 0–07–062580–8

CONTENTS

	Preface	ix
Chapter 1	Linear Programming—Origin and Definition	1
1-1	Origin	1
1-2	Graphical Illustrations—Maximization	2
1-3	Graphical Solution—Alternate Optima	5
1-4	Graphical Solution—Minimization	6
1-5	Graphical Solution—Unbounded Feasible Region	8
	Exercises	11
Chapter 2	Basic Concepts of Linear Algebra	13
2-1	Introduction and Illustrative Problem	13
2-2	Primal Problem	15
2-3	Dual Linear Programming Problem	16
2-4	Matrices	18
2-5	Examples of Simple Matrix Operations	20
2-6	Scalar Product of Two Vectors	22
2-7	Multiplication of Matrices	24
2-8	Linear Programming in Matrix Notation	25
2-9	Identity, Null, and Transposed Matrices	28
2-10	Matrix Inverse	31
2-11	Algebraic Solution of a Linear System of Equations	33
2-12	Finding the Inverse of a Matrix	35
2-13	Euclidean Space and Linear Independence	36
2-14	Special Properties of Linear Independence	39
2-15	Bases and Spanning Sets of Vectors	40

2-16 Replacement of a Basic Vector with an Arbitrary Vector 42
2-17 Number of Vectors in a Basis for E^n 44
2-18 Vector Spaces and Subspaces 44
2-19 Rank of a Matrix 44
2-20 Product Form of the Inverse of a Matrix 45
2-21 Illustrative Example of an Inverse 47
2-22 Linear Systems of Equations 49
2-23 Basic Solutions 49
2-24 Number of Basic Solutions 50
2-25 Degeneracy and Feasibility 51
2-26 Point Sets 51
2-27 Straight Lines and Hyperplanes 51
2-28 Convex Sets 52
2-29 Linear Programming—Concept of a Solution 53
2-30 Matrix Example 54
Exercises 54

Chapter 3 Simplex Method and Interpretation of the Tableaux 58

3-1 Introduction to Simplex Method—Matrix Statement 58
3-2 Basic Feasible Solutions 60
3-3 Nonbasic Variables as a Linear Combination of Basic Variables 62
3-4 Selection of Incoming Vector 64
3-5 Selection of Outgoing Vector 64
3-6 Second Iteration 67
3-7 Third Iteration 68
3-8 Fourth Iteration 69
3-9 Comments 71
3-10 Computer Solution 72
3-11 Comments on Linear Independence 72
3-12 Applications 74
3-13 Use of the Objective Function as an Additional Constraint 76
3-14 Alternative Logic 79
3-15 Basic Inverse 79
3-16 Summary 80
Exercises 81

Chapter 4 Generalized Linear Programming Problems 83

4-1 Equal-to-or-Greater-Than and Equal-to-Constraints 83
4-2 Use of Artificial Variables 84
4-3 Meaning of the Artificial Variables 88
4-4 Two-Phase Method of Handling Artificial Variables 89
4-5 Minimization Problems 92
4-6 Solution Example Involving Minimization and a Mixture of Constraints 93
4-7 Comments on General Solutions 95
4-8 Application in Steel Fabricating 96
Exercises 98

Chapter 5 Sensitivity Analysis 102

5-1 Problem Modification to Test Sensitivity 102
5-2 Duality 102
5-3 Some Fundamental Properties of the Primal and Dual Problems 104
5-4 Computations 107
5-5 Simplex Solution for General Problems 108
5-6 Dual Values Corresponding to Equal-to-or-Less-Than Constraints 108
5-7 Dual Values Corresponding to Equal-to-or-Greater-Than Constraints 109
5-8 Dual Values Corresponding to Equal-to Constraints 109
5-9 Summary of Dual Solutions 110
5-10 Process for Writing the Dual Statement of a Problem 110
5-11 Changes in the Components of the Right-Hand Side 112
5-12 Application 117
5-13 Changes in the Coefficients of the Objective Function 118
5-14 Addition of a New Variable 120
5-15 Changes in the Elements of Matrix **A** 121
5-16 Addition of a New Constraint 123
5-17 Removal of a Constraint 125
5-18 Summary on Sensitivity Analysis 127
Exercises 128

Chapter 6 Revised Simplex Procedure 130

6-1 Introduction 130
6-2 Revised Tableau 131
6-3 Selection of Incoming Variable 132
6-4 Selection of Outgoing Variable 133
6-5 Updated Basis Inverse 134
6-6 Condensed Tableau 135
6-7 Condensed Second Tableau 135
6-8 Condensed Third Tableau 137
6-9 Condensed Fourth Tableau 138
6-10 Condensed Optimal Tableau 140
6-11 Condensed Summary of Tableaux 141
6-12 Summary of Revised Simplex Method 142
Exercises 143

Chapter 7 The Transportation Model 144

7-1 Introduction to the Transportation Model 144
7-2 Special Form of Problem—Dual Statement 145
7-3 Development of Solution Routine 147
7-4 Degeneracies in the Solution 152
7-5 Generalizing the Transportation Model 154
7-6 Efficient First Feasible Selection 155
7-7 Assignment Problem 158
Exercises 158

Chapter 8 Computer-Generated Information 161

8-1 Use of the Computer 161
8-2 Scheduling Problem 161
8-3 Solution—Nature of the Results 166
8-4 Solution Results—Actual Solution 166
8-5 Solution Results—Opportunity Costs 168
8-6 Alternate Optimal Solutions 169
8-7 Sensitivity on the Right-Hand Side 170
8-8 Sensitivity on Coefficients of Objective Function 170
8-9 Optimal Solution—Basis Inverse 173
8-10 Computer Codes for Linear Programming 173
Exercises 173

Chapter 9 Importance of Modeling for Interpretation of Linear Programming Problems 180

9-1 Introduction 180
9-2 Scheduling Problem 180
9-3 Initial Intuitive Model Formulation 182
9-4 Mathematical Statement of the Problem 183
9-5 Optimal Solution 185
9-6 Efficiency of the Model 186
9-7 Alternative Statement of the Scheduling Problem 188
9-8 Scheduling Parameters 190
9-9 Advantages of the Second Model 191
9-10 Sensitivity on Recovery Rates 191
9-11 Sensitivity on Processing Rates 196
9-12 Sensitivity on Processing Costs 197
9-13 Effects on the Computer 197
9-14 Criteria for Modeling 200
9-15 Alternative Formulation with a Reduced Model 201
9-16 Solution Information from the Reduced Model 203
9-17 Efficiency of Linear Programming Models 206
9-18 Additional Model Characteristics 207
9-19 Goal Programming 208
Exercises 212

Index 215

PREFACE

This text has been written primarily for those college students who have no previous background in linear programming and are planning careers as administrators or managers in business, nonprofit organizations, or government. The text can be used in its entirety in either a semester or two-quarter course at the advanced undergraduate or early graduate level in business administration, economics, or engineering. A sound background in college algebra is necessary; any further mathematics that is needed is provided within the text.

OBJECTIVES

The principal objectives are

* To introduce the important ideas in linear programming at a fundamental level
* To assist students in understanding a wide range of linear programming models
* To motivate students to continue their study of linear programming applications
* To provide students with the fundamentals of computer use to solve linear programming problems
* To assist the student in the formulation, solution, and interpretation of linear programming models

COVERAGE AND EMPHASIS

A course in linear programming can be given at various levels, ranging from the cookbook variety, with little understanding of the theory, to the highly theoretical, with little understanding of the practical aspects. This text attempts to follow neither of these extremes, but to include the best features of each.

I am a firm believer in using the computer to solve linear programming problems. Students must therefore become quite familiar with particular linear programming codes. I have found that early introduction of computer methods enhances the overall success of the course. My philosophy is to present the ideas in matrix notation, to give a fuller understanding of what actually happens in the process of obtaining a solution to a problem.

Throughout the text I have attempted to use as small a model as possible to illustrate the principles being covered while solving these problems by manual methods. Consequently, it is difficult to approach real-world problems until you get to the chapters on modeling and computer solutions. Some of the early chapters do call for the setting up of problems without actually carrying out the solutions.

It is important to cover the topics in sequence rather than to skip around in the text. Where readers are knowledgeable on certain topics, they should still cover enough of them to maintain the thread of ideas that weaves through the text.

The book is organized in the following fashion. Chapter 1 introduces the linear programming concept by presenting a number of problems. These problems are solved by the graphical method in order to produce a mental image of the solution process together with difficulties introduced whenever more than two variables appear in the original statement of the problem.

Chapter 2 introduces a specific maximization problem with all equal-to-or-less-than constraints and begins its solution by means of some basic algebra. The solution progresses until the algebra becomes burdensome and gives some indication that a good systematic way of handling the problem is desirable. (The same example is used later as a vehicle to present other ideas.) It is then stated that matrix algebra provides the basis for this system, and the remainder of the chapter is devoted to basic concepts of matrix algebra. The principles developed lead to the solution of linear systems of equations by matrix methods without the burden of dealing with situations in which the matrix of coefficients is of less than full rank. The concepts of linear independence of vectors, basis set of vectors, and spanning set of vectors are developed. These are then applied to systems of linear equations in order to discuss basic solutions to linear systems in terms of basic and nonbasic variables. The material concentrates on the idea that one should be able to recognize independent sets of vectors and basic solutions to linear systems of equations without placing emphasis on either redundant or inconsistent systems, since these offer no great difficulty in the solution of linear programming problems.

Chapter 3 returns to the sample problem started at the beginning of Chapter 2 and uses matrix algebra developed earlier to derive the simplex solution. Detailed computations are given to make clear such concepts as slack variables, the test for optimality, determination of entering and leaving variables, transformations to shift from one basic solution to another, the inverse of a basis set of vectors, and the use of the objective function as just another constraint. Thorough understanding of this chapter should enable one to solve in detail any linear programming problem whose solution involves maximization with equal-to-or-less-than constraints.

Chapter 4 extends the ideas of Chapter 3 to include problems with equal-to-or-greater-than and equal-to constraints by introducing negative slack or surplus and artificial variables when appropriate. In addition, minimization problems are handled by changing such problems into maximization problems, solving, and returning the solution to results applicable to the original problem. In order to handle the artificial variables, both the big-M method and the two-phase method are described and analyzed, with detailed tableaux given for each sample problem. At this point, such concepts as redundancy, inconsistency, and unbounded solutions are discussed, but not in theoretical detail.

Chapter 5 deals with sensitivity analysis resulting from changes in the $\mathbf{b}$ vector, changes in the $\mathbf{c}$ vector, addition of a new activity level, changes in the coefficients of the constraints, addition of a new constraint, and the removal of a constraint—all applied to an already optimal solution. In order to study these concepts, the principle of duality is first presented and three very basic theorems are proved using matrix algebra. The results of these theorems, together with the matrix algebra previously developed, are then used for the sensitivity analysis. Much computational detail is shown for the sample problems of Chapters 3 and 4, to make the concept of duality clear.

Chapter 6 develops the basic concepts of the revised simplex method, since this is so important for many computational methods, in particular for problems handled by the decomposition principle. Details are given for examples used previously.

Chapter 7 discusses and develops the method of solution of the transportation model. The principle of duality is used to develop the algorithm.

Chapter 8 discusses use of the computer for the solution of linear programming problems. It is discussed with reference to a particular program, MPOS, developed at Northwestern University. Even though the discussion gives information pertaining to MPOS, it is sufficiently general that one should be able to apply the methods to any linear programming code that might be available. Many details are given for the interpretation of the output, and it is related particularly to the material on sensitivity analysis given in Chapter 5.

Chapter 9 emphasizes the importance of proper modeling of a linear programming problem to an in-depth understanding of its solution. Since the principles involved have an element of heuristics, the material in this chapter has not been stated in the form of an algorithm. Two very enlightening examples are discussed in great detail in order to emphasize modeling.

The text can be used for courses of varying length. Some of the material could be omitted if the course is no more than one quarter in length. The first five chapters are essential for a course of any length, and the remaining chapters add considerable depth of understanding. Chapter 6 could be omitted if no further work in linear programming is contemplated. Chapter 7 is particularly important for readers who are planning to study production or operations management. Chapters 8 and 9 add conceptionally important material to the first five chapters. In any case, the first five chapters should be studied as a unit, and after that, any of the next four chapters can be studied independently in any order.

DEVELOPMENT OF THE TEXT

The material as presented has been used in draft form in several classes taught by the author during the last three years. It has gone through several revisions during that time and has been very well received by students. Although matrix algebra concepts are new to a great majority of students studying linear programming for the first time, the concepts have been understood almost without exception. I have found unusual success in teaching linear programming through the use of matrix algebra and strongly feel that it is the only way in which it should be done if one is to get any depth of understanding.

ACKNOWLEDGMENTS

My thanks are due to many of my colleagues at Northwestern University for their valuable comments during our many discussions concerning the way in which I have taught the material on linear programming.

Particular thanks are due Professor Martin Starr of Columbia University for his critical review of the manuscript. He was able to offer many suggestions that have been incorporated without changing my basic philosophy.

To our daughter, Sue, I express thanks for her suggestions for getting the material across to her generation.

To my wife, Winifred, I express particular thanks and appreciation for encouraging me and for allowing me to use so much of the time that was rightfully hers. She has experienced many lonesome moments while I was laboring over many details, torn between getting the text into print and enjoying her company.

Any errors that exist are solely mine. I trust that there will be few.

Leonard W. Swanson

LINEAR PROGRAMMING
Basic Theory and Applications

CHAPTER

ONE

LINEAR PROGRAMMING ORIGIN AND DEFINITION

1-1 ORIGIN

The methods of linear programming are essentially new, having been developed during the last 30 years. Although many references can be made to work along this general line, the initial work was developed by George Dantzig in about 1947. George Dantzig was a member of an Air Force Group that developed the simplex method of solution, but his work was not generally available until 1951. Even though the theory had been devised before 1950, very little use was made of it until the computer came into being, since the computer is so essential to the solution of problems of even moderate size.

The basic problem of linear programming is one of either maximizing or minimizing a function of several variables, with the variables being subject to a number of constraints. The constraints, as well as the function being optimized, are linear. Although calculus is helpful in many types of optimization problems, it is of no value in this type of problem—hence the necessity for developing an entirely new procedure.

In the development of the material that follows, we shall be concerned with the theory as well as the techniques for the solution of linear programming problems, but a great deal of emphasis will be placed upon the practical approach to the solution of problems and to a full understanding of its important aspects. The general linear programming problem deals with the allocation of resources in seeking the optimization. The fact that the resources are limited means that a choice must be made in assigning the use of these resources. With only a few constraints, the problem becomes complicated enough that intuitional choices or those based upon experience alone are seldom correct.

1-2 GRAPHICAL ILLUSTRATIONS—MAXIMIZATION

To introduce linear programming, we shall use a few graphical illustrations. Since it is easy to find graphical solutions and give geometric interpretations in two dimensions, these illustrations will be limited to two dimensions. Initially these examples will be elementary.

Example 1-1 Jim and Jane have a small furniture workshop in their garage, where they assemble from purchased parts and finish the furniture in preparation for sale. They are presently limiting their production to tables and chairs. Each chair requires 4 h to assemble and 2 h to finish, but each table requires 2 h to assemble and 4 h to finish. Jim works only on the assembly operation, Jane works only on the finishing operation, and each puts in no more than an 8-h day. It is known that each chair can be sold for $300 and each table for $200. What should the average daily production of tables and chairs be to maximize the total income?

Mathematical Model

To obtain a solution to this problem, it is necessary first to write a mathematical model of the problem. Some simplifying assumptions are made to do this, but one must keep in mind that it is necessary to make realistic assumptions if the solution is to have any meaning. For this solution, let

x_1 = number of chairs to be produced
x_2 = number of tables to be produced

Consider the income to be measured in units of *hundreds of dollars*, so that the total income to be maximized will be given by

$$3x_1 + 2x_2$$

Note that the income on chairs is directly proportional to the number being produced and that the income on tables is directly proportional to the number being produced; the income to be maximized is a *linear* function of the number of chairs being produced and the number of tables being produced. This linear property is one of the basic tenets of linear programming.

Since each chair requires 4 h for assembly and each table requires 2 h for assembly, the total amount of time in the assembly operation is given by

$$4x_1 + 2x_2$$

Since assembly time is limited to 8 h, we have the constraint

$$4x_1 + 2x_2 \leqq 8$$

Since each chair requires 2 h of finishing and each table requires 4 h of finishing, the total amount of time in the finishing operation is

$$2x_1 + 4x_2$$

and since Jane performs the finishing task during her day, we have the constraint

$$2x_1 + 4x_2 \leqq 8$$

With the further restriction that negative solution values are not allowed, the mathematical model becomes:

Find

$$x_1, x_2 \geqq 0$$

such that

$$4x_1 + 2x_2 \leqq 8$$

$$2x_1 + 4x_2 \leqq 8$$

and such that

$$z = 3x_1 + 2x_2 \text{ is a } \textit{maximum}$$

To solve this problem graphically, we first construct the set of points (x_1, x_2) which represents a feasible solution to the problem. This is the set of points that satisfies all the constraints. This set of points, called the *solution space*, is shown by the shaded section in Fig. 1-1.

The solution space is bounded by four straight lines and lies entirely within the first quadrant. This is explained by the fact that the restriction

$$4x_1 + 2x_2 \leqq 8$$

requires that all points be either on or below the line

$$4x_1 + 2x_2 = 8$$

Similarly, the restriction

$$2x_1 + 4x_2 \leqq 8$$

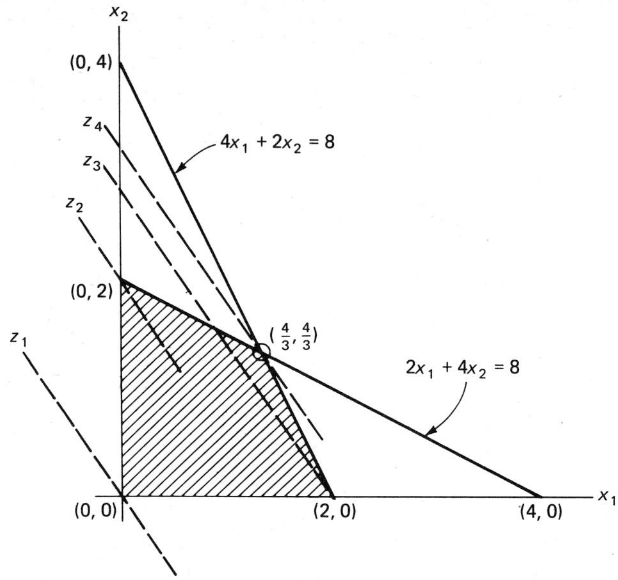

Figure 1-1

requires that all points be on or below the line

$$2x_1 + 4x_2 = 8$$

The constraints

$$x_1, x_2 \geqq 0$$

restrict the points to the first quadrant.

For any selected value of z the function

$$z = 3x_1 + 2x_2$$

is a straight line. For each value of z we have a different straight line, but all the lines in the system are parallel simply because the slope of any one of these lines is completely independent of the value of z.

We wish to find the line from the entire system of parallel lines with the largest value of z while having at least one point in or on the boundary of the solution space. On the graph:

z_1 passes through (0, 0) and has the value zero.
z_2 passes through (0, 2) and has the value 4.
z_3 passes through (2, 0) and has the value 6.
z_4 passes through $(\frac{4}{3}, \frac{4}{3})$ and has the value $\frac{20}{3}$.

It is evident that the value of z increases as the line is moved farther and farther away from the origin. Thus for the two-dimensional case it is necessary first to determine the direction of the line representing the function to be maximized. Then the line is moved as far away from the origin as possible while keeping at least one point in the feasible region. For this case it is easily seen that the point $(\frac{4}{3}, \frac{4}{3})$ is the only point meeting these conditions and hence one needs only to evaluate z for this point to find the maximum value.

Interpretation of the Solution

Consider next the meaning of the optimal solution just found. It shows that Jim and Jane should make $\frac{4}{3}$ chairs and $\frac{4}{3}$ tables each day, with a resulting income of $\frac{20}{3}$ hundreds of dollars, or $667 per day. You might say that one does not make one and one-third chairs or one and one-third tables, but considering that one would have essentially a continuous production line, 4 chairs and 4 tables could be produced every 3 days, so the results merely tell us that Jim and Jane make equal numbers of chairs and tables and that the rate of production is $1\frac{1}{3}$ of each per day.

At this point you might wonder whether there exists a scheduling problem related to making sure that there is always something assembled for Jane to finish and always a supply of parts for Jim to assemble. Actually, at this stage we ignore that problem by suggesting that we are interested here only in the product-mix problem, so that we always assume we have a modest stockpile of parts on hand and in turn have a few tables and chairs assembled so that Jim and Jane both have work to do. With these assumptions we can concentrate on the product-mix problem, as we have done above. Note that the solution is *unique*.

1-3 GRAPHICAL SOLUTION—ALTERNATE OPTIMA

Example 1-2 Suppose now that we modify the income for Example 1-1 so that each chair will bring $200 and each table will bring $100. All other facets of the problem will remain the same.

Mathematical Model

The function to be maximized is now

$$z = 2x_1 + x_2$$

With all constraints remaining as before so that the feasible solution space is unaltered, Fig. 1-2 shows that the line z is parallel to one of the boundaries of the solution space. That boundary is marked PQ.

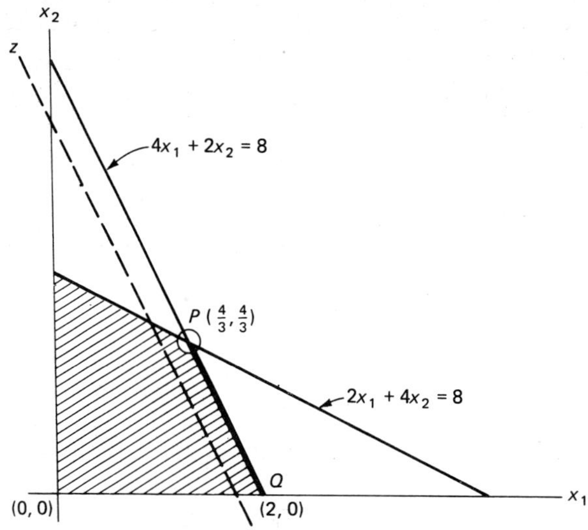

Figure 1-2

In this case, in moving the line z as far from the origin as possible while keeping at least one point from the feasible region on the line, it is found that all the points on the line segment PQ qualify. The point $x_1 = \frac{4}{3}$, $x_2 = \frac{4}{3}$ produces an optimal solution value of $z = \$400$. Similarly, $x_1 = 2$, $x_2 = 0$ produces an optimal income of \$400, so the solution is *not* unique. There are, in fact, an infinite number of solutions for x_1 and x_2 that produce the optimal income of \$400. The managerial implications of this are that one must choose the best solution influenced by factors other than those included in the linear programming formulation. For example, one might want to make $x_1 = x_2$ in order to have a balanced product line. This is acceptable, since there is a point on the line PQ where $x_1 = x_2$.

1-4 GRAPHICAL SOLUTION—MINIMIZATION

Example 1-3 Let us now consider Jim and Jane's operation in Examples 1-1 and 1-2 from a slightly different point of view. Income no longer seems to be of primary concern, since there are a number of fixed-cost operations. There is a variable cost of \$30 for each chair and \$20 for each table. We now wish to minimize the variable cost subject to a number of different types of restrictions. The sale of chairs cannot exceed *three units* and the sale of tables cannot exceed *four units*; furthermore, the sale of chairs and tables together cannot exceed *five units.* With the new approach, a drying room is used after the gluing has been completed. A chair requires 1 h and a table requires 5 h in the drying room. Because startup is costly, operating the room is very expensive unless it is used for at least 5 h each day. There are also requirements with regard to packaging. Five machine runs are required for a chair and two for a table. It is impractical to set up this machine unless at least 10 runs are to be made.

Mathematical Model

Let

x_1 = number of chairs produced
x_2 = number of tables produced

Then the model is:

Find

$$x_1, x_2 \geqq 0$$

such that

$$\begin{aligned} x_1 \qquad &\leqq 3 \\ x_2 &\leqq 4 \\ x_1 + x_2 &\leqq 5 \\ x_1 + 5x_2 &\geqq 5 \qquad \text{(drying-room hours)} \\ 5x_1 + 2x_2 &\geqq 10 \qquad \text{(packaging runs)} \end{aligned}$$

and such that

$$Z = 3x_1 + 2x_2 \text{ is a } \textit{minimum}$$

For simplicity, the results for Z are given in units of \$10. The feasible solution space, together with the solution, is given in Fig. 1-3.

For this problem it is necessary to place the Z line as close to the origin (minimization) as possible while keeping at least one point of the feasible region on the line. As shown in Fig. 1-3, the feasible region is bounded by five straight lines. The Z-line direction is shown by the dashed line through the origin and the minimizing line must pass through the point $(\frac{40}{23}, \frac{15}{23})$, which gives the minimum value of $\frac{150}{23}$. The solution is unique, as in Example 1-1. The point that gives the minimum is found by determining the intersection of the lines

$$\begin{aligned} x_1 + 5x_2 &= 5 \\ 5x_1 + 2x_2 &= 10 \end{aligned}$$

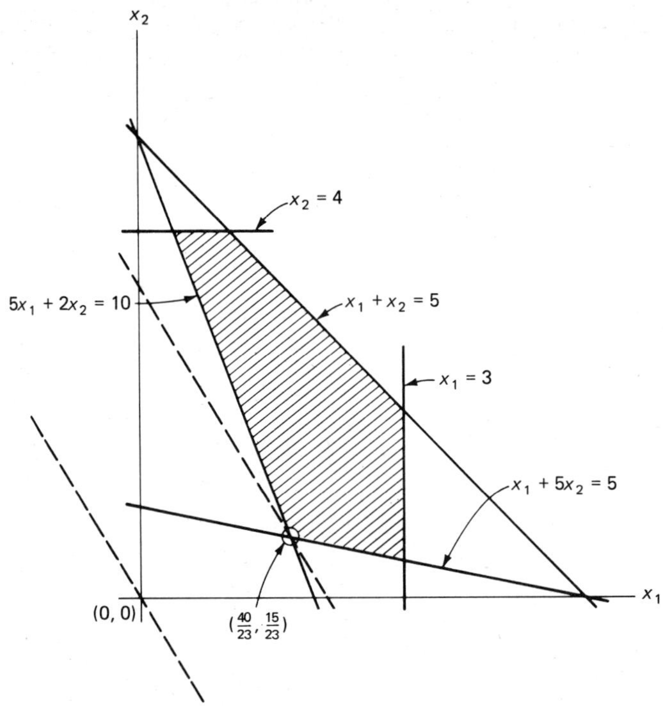

Figure 1-3

1-5 GRAPHICAL SOLUTION—UNBOUNDED FEASIBLE REGION

The examples used previously are fairly straightforward. With different kinds of constraints, some rather unusual situations evolve. Consider the following example.

Example 1-4 Suppose that you are faced with a decision involving the buying of shares of stock in two small, closely held companies. Because of your friendship with the management of these two companies, you are really not free to make your selections on a cold, calculating basis. You have imposed some restrictions upon yourself and you are bound to some other restrictions. Without reference to price, you are obligated to hold a number of shares in Jones's company that is at least equal to half of the number of shares in Smith's company. Furthermore, the number of shares in Jones's company must be no more than 400 more than that for Smith's company. The current situation is that by holding Smith's stock, you lose about \$2 a share and by holding Jones's stock you gain about \$1 per share. If you are interested in maximizing your gains in holding these two stocks, how many shares of each should you be holding? It appears that under the circumstances, you would really not hold any stock in Smith's company, but for other reasons a requirement was made that you hold at least 100 shares.

Mathematical Model

Let

x_1 = number of shares in Smith's company (measured in hundreds)
x_2 = number of shares in Jones's company (measured in hundreds)

The net result of the restrictions, taken in the order in which they were mentioned, yields

$$x_2 \geqq \tfrac{1}{2}x_1$$

$$x_2 \leqq 4 + x_1$$

$$x_1 \geqq 1$$

with $$z = -200x_1 + 100x_2 \text{ to be } \textit{maximized}$$

With proper simplification of the constraints as given, the more formal model is:

Find

$$x_1, x_2 \geqq 0$$

such that

$$\begin{aligned} -x_1 + 2x_2 &\geqq 0 \\ -x_1 + x_2 &\leqq 4 \\ x_1 &\geqq 1 \end{aligned}$$

and such that

$$z = -200x_1 + 100x_2 \text{ is a } \textit{maximum}$$

Figure 1-4 shows the solution space; it should be noted that the solution space is unbounded. The direction of the objective function is shown by the dashed line through the origin. It can be seen that by moving the z line to the right, the value of z is diminished; by moving it to the left, the value of z is increased. To find the maximum value of z, the line must be moved as far to the left as possible while keeping at least one point of the feasible region on the line. At the point where the z line passes through (1, 5), z takes on its maximum value of 300. This case is rather unusual, since it is seen that the

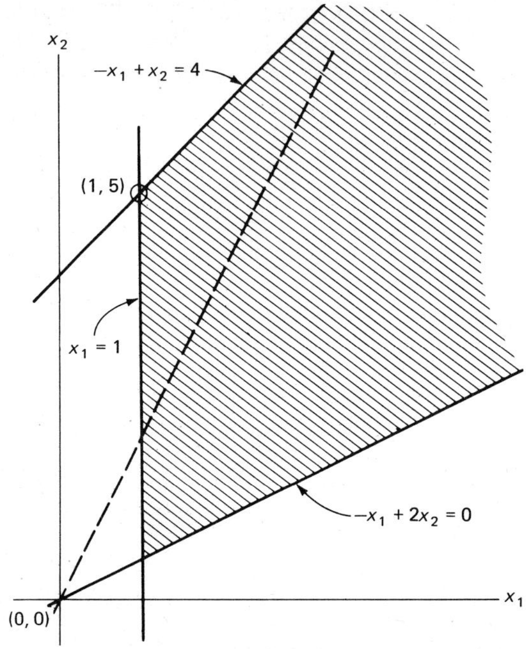

Figure 1-4

feasible region is *unbounded* and yet a maximum value occurs for the objective function. Interesting? The result is unique and the values of both x_1 and x_2 are finite.

Had we tried to find the minimum value of z, we would find that none exists; that is, we can continue to move the line to the right without reaching any limiting position, and consequently z becomes smaller and smaller, without limit.

Consider for this same feasible region that you lose only \$1 per share on Smith's stock, so that the objective function becomes

$$z = -100x_1 + 100x_2$$

In this case we find that when the z line is as far to the left as it will go, it will not only pass through the point (1, 5) with a value of 400, but it coincides with the constraining line,

$$-x_1 + x_2 = 4$$

so that many pairs of values of x_1 and x_2 give the same optimal value for z. For example, $x_1 = 16$ and $x_2 = 20$ yields a value of 400 for z; similarly, any pair of values of x_1 and x_2 such that x_2 is 4 units larger than x_1 will provide the maximum value of 400 for z. This case is particularly interesting because there is no upper limit to the values of either x_1 or x_2 within the feasible region, yet a maximum value of the objective function does exist.

The examples in this chapter give some indication of the kinds of situations that may arise in the solution of linear programming problems. Whatever analytical methods are developed for the solution of the general linear programming problem, each of the preceding situations should be identified in any sophisticated method of solution. The illustrations are not intended to represent all the situations that might occur, but they are representative of the fact that there are many variations from a standard simple problem. Chapter 2 introduces some of the specifics of linear programming by means of examples.

EXERCISES

The exercises that follow will be more instructive if they are solved in the sequence in which they are given. It is intended that some of the principles involved in a given exercise will be used in later exercises.

1 A small electronics manufacturer produces two models of hand calculators, the business calculator and the scientific calculator. Each of these types of calculators goes through three departments: wiring, assembly, and testing. The manufacturer estimates that there is a $4 profit on the business calculator and a $10 profit on the scientific calculator. The following information is known about the production process:

Each business calculator requires:

1 h in wiring
3 h in assembly
1 h in testing

Each scientific calculator requires:

4 h in wiring
1 h in assembly
1 h in testing

For the scheduling period, the following hours are available:

24 h in wiring
21 h in assembly
9 h in testing

Find out how many of each type of calculator should be made to maximize profits.

(*a*) Write the mathematical statement of the problem that must be solved.

(*b*) Find the solution by using the graphical method.

2 A farmer feeds his pigs with an all-purpose grain and a dietary supplement. The all-purpose grain costs $10 per bag and the supplement costs $15 per bag. The grain provides 2 calorie units and 1 vitamin unit in each bag, while the supplement provides 1 calorie unit and 4 vitamin units in each bag. It has been decided that to maintain a proper growth pattern, the pigs must be furnished a minimum of 200 calorie units and 160 vitamin units per day. What is the minimum daily food cost of this program, and how much of each ingredient should be purchased? Solve by the graphical method.

3 The Sound-Off Company produces black-and-white TV sets and a self-contained stereo set. The per unit profit for the TV set is \$15 and the per unit profit for the stereo set is \$45. The schedule for production of these two items is made up on a weekly basis. There are a number of restrictions as follows:

1. Special speakers are required for the stereo set and production is limited to 50 sets/week.
2. The preliminary assembly of a TV set requires 6 h per set and the preliminary assembly of a stereo set requires 18 h per set. The department performing these operations has 1800 labor-hours available weekly.
3. Once the preliminary assembly is made, each set is mounted in a cabinet. The cabinet assembly of a TV set requires 1 h and the stereo set requires 1.6 h. There are 240 labor-hours available each week in this department.
4. After the sets are assembled, inspection takes place. Each TV set requires $\frac{1}{2}$ h for inspection and each stereo set requires 2 h for inspection. There are a total of 162 h per week available for inspection.

By using the graphical method of solution, determine the optimal mix of production of TV and stereo sets to maximize profit.

(*a*) Is the solution unique?

(*b*) Could any of the constraints be removed without affecting the solution of the problem; that is, are any of the constraints redundant? Which ones?

(*c*) Within what range could the profit on the TV set vary from the present \$15 without changing the optimal mix of production? What is the range of profit with this range?

4 The Minute Housewares Company makes sweepers and polishers. Each sweeper has a profit of \$1. Each polisher has a profit of either 50 cents (case A) or \$1.50 (case B).

Each sweeper requires 1 labor-hour and one run on a special machine.
Each polisher requires 1 labor-hour and two runs on the special machine.
For the scheduling period, the demand is such that at least 10 sweepers must be made and at most 5 polishers.
There are 20 labor-hours and 20 machine runs available.

Use the graphical method of solution to find the *maximum* profit for the scheduling period under each of the following conditions:

(*a*) The profit on the polisher is 50 cents.

(*b*) The profit on the polisher is \$1.50.

In an effort to increase profits, it was found that an additional five runs on the special machine could be rented for 25 cents each. Should you rent the additional five runs? Answer for each of the conditions below.

(*c*) The profit on the polisher is 50 cents.

(*d*) The profit on the polisher is \$1.50.

(*e*) If you could rent more than five extra runs in either of the cases above, would you? What is the optimal number of extra runs to be rented?

CHAPTER

TWO

BASIC CONCEPTS OF LINEAR ALGEBRA

2-1 INTRODUCTION AND ILLUSTRATIVE PROBLEM

We have seen how to find the optimal solution to a two-variable linear programming problem by the graphical method. Such a method is never practical for problems with more than two variables, and since most real-world problems contain more than two decision variables, we do not expect to use the graphical procedure for the solution of real-world problems. The method was introduced in Chap. 1 to give the reader a visual introduction to some of the complexities and interesting features of linear programming problems. It was never intended that the graphical procedure be used as a regular decision-making process. An algebraic procedure will be used to solve these real-world problems. The background for this algebraic procedure will be developed in this chapter and the method itself will be developed in Chap. 3. Once the procedure has been formulated into an algorithm, we shall rely on the computer to carry out the arithmetic computations.

Consider now a fairly typical situation faced by a manufacturer who has several hundred people working for her in the production of four metal product items that she is able to manufacture with the equipment and personnel that she has available. The four different products are four different models of a specialized housing which is supplied to the electronic industry.

We shall call the models P, Q, R, and S. The conditions are such that she can actually sell all that she can make, so it is up to her to decide what the product mix should be.

Even though many operations are required in the production of each of the models, the most critical ones are in machine centers A, B, and C. Machine center A is a stamping and punch-press operation, machine center B is a drilling operation, and machine center C is a finishing and polishing operation. Each of the products must go through each of the machine centers with varying amounts of time required. Machine center A has 2000 h available for the scheduling period, machine center B has 8000 h, and machine center C has 5000 h. Model P will generate \$5.24 in profit for each unit produced, model Q will generate \$7.30, model R will generate \$8.34, and model S will generate \$4.18. Table 2-1 summarizes this information and also shows the time required for each of the products in each of the machine centers.

From Table 2-1 it is seen that product P requires 1.5 h in machine center A, 1 h in machine center B, and 1.5 h in machine center C. For each unit manufactured there is a profit of \$5.24. Similar information can be found for each of the other products. What mix of products should be made to maximize profits?

To deal with the problem as a product-mix problem rather than a scheduling problem, we shall assume that each product must go through centers A, B, and C in that order but that there are always a sufficient number of partially finished products in inventory so that no machine center ever has to wait for material on which to work. This is really not an impractical assumption, since we think of the production lines as being continuous and the cost of the in-process inventory as being small.

Only slight investigation reveals that manufacturing all you can of product R will not yield the maximum profit, even though it has the highest per unit profit. The reasoning is as follows. The hourly restriction in machine center A restricts the number of units of R that can be manufactured to

Table 2-1

Machine center	Products				Total time available
	P	Q	R	S	
A	1.5	1	2.4	1	2000
B	1	5	1	3.5	8000
C	1.5	3	3.5	1	5000
Profit per unit	\$5.24	\$7.30	\$8.34	\$4.18	

2000/2.4 = 833.33 units without using the full facilities of either machine center *B* or *C*, and yet there would be no time available in machine center *A* to manufacture any of the other products.

By reducing the number of units of *R*, facilities would be made available for the manufacture of *P*, *Q*, or *S* with additional profits. For example, if the quantity of *R* to be manufactured is reduced by 1 unit, 2.4 h are freed in machine center *A*, and this allows the manufacture of 2.4 units of product *Q*, each of which contributes a profit of $7.30 with a net gain of $9.18 (i.e., a gain of 2.4 times $7.30 and a corresponding loss of *one* times $8.34). This is possible only because of the idle hours in machine centers *B* and *C*. This rate of gain will not continue for every unit by which you cut back the production of *R*, but it will continue for a substantial number of units.

If one is to find the optimal product mix, many analyses like the one above must be made. With many products, the analyses become very complex, so that a routine must be developed. The routine is called *linear programming*. The term "linear" is applied since the resources used are in direct proportion to the number of units being manufactured; similarly, the profits are directly proportional to the number of units being manufactured. The linearity property is assumed to be true, but such an assumption often does not seem to be out of line.

The method of solution of linear programming problems has its foundation in the theory of solution of linear systems of equations, which in turn is dependent upon matrix algebra for a fundamental understanding. For the problem that we have introduced, let us formulate a mathematical model. To do this, let

x_1 = number of units of product *P* to be manufactured
x_2 = number of units of product *Q* to be manufactured
x_3 = number of units of product *R* to be manufactured
x_4 = number of units of product *S* to be manufactured

Since you cannot manufacture a negative quantity, these variables must be nonnegative.

2-2 PRIMAL PROBLEM

Except for rather simple problems, it is not likely that two persons would model a problem in exactly the same way, even though each may have a correct model. In order that we may make reference to the model, we shall designate that initial model for the problem as the *primal problem*. For our problem in Sec. 2-1, the primal problem is not difficult to formulate. There is a constraint on hours available for each of the machine centers. The model can then be stated as follows:

Find

$$x_1, x_2, x_3, x_4 \geqq 0$$

such that

$$\begin{aligned} 1.5x_1 + x_2 + 2.4x_3 + x_4 &\leqq 2000 \\ x_1 + 5x_2 + x_3 + 3.5x_4 &\leqq 8000 \\ 1.5x_1 + 3x_2 + 3.5x_3 + x_4 &\leqq 5000 \end{aligned} \tag{2-1}$$

and such that

$$z = 5.24x_1 + 7.30x_2 + 8.34x_3 + 4.18x_4 \text{ is a } maximum$$

A solution to this mathematical problem yields the optimal product mix for the scheduling period.

2-3 DUAL LINEAR PROGRAMMING PROBLEM

The mathematical solution to problem (2-1) formulated in Sec. 2-2, is straightforward and in itself is not very exciting, as will be shown later. If we realize, however, that perhaps every one of the numerical quantities appearing in the model is not necessarily accurate, we begin to ask ourselves some questions. How would the solution change if we make more hours available in machine center *A* or *B* or *C*, or how would the solution change if one or more of the per unit profits should change? What if we change some of the coefficients in a constraint? What happens if we add a constraint or remove a constraint? We see that if we solve the original problem and then answer a series of questions similar to those above, we are really solving a whole series of related problems.

One of the natural questions that might be asked is: What would an additional hour in machine center *A* be worth? We might be interested in more than one additional hour, so suppose we let y_1 represent the value of each additional hour and think of it as having the units of rate change of profit per unit change in the 2000 h that are presently available in machine center *A* (i.e., the units are dollars per hour). We shall prove later that this definition is appropriate, legitimate, and particularly meaningful. Another definition for y_1 is that it represents the imputed hourly cost of running machine center *A* when the entire operation is scheduled optimally.

This ultimately leads us to the formulation of another linear programming problem. At this stage, let

y_1 = marginal value of an hour in machine center *A*
y_2 = marginal value of an hour in machine center *B*
y_3 = marginal value of an hour in machine center *C*

You will note that the variables defined are concerned with the values of additional units of the resources as expressed on the right-hand sides of the constraints for the primal problem. With these definitions, then, consider the following linear programming problem:

Find

$$y_1, y_2, y_3 \geqq 0$$

such that

$$\begin{aligned} 1.5y_1 + \quad y_2 + 1.5y_3 &\geqq 5.24 \\ y_1 + \ 5y_2 + \quad 3y_3 &\geqq 7.30 \\ 2.4y_1 + \quad y_2 + 3.5y_3 &\geqq 8.34 \\ y_1 + 3.5y_2 + \quad y_3 &\geqq 4.18 \end{aligned} \tag{2-2}$$

and such that

$$Z = 2000y_1 + 8000y_2 + 5000y_3 \text{ is a } minimum$$

At this point it is difficult to understand how one might have thought to even pose this problem, but it so happens that it has a very important relationship to primal problem (2-1) in Sec. 2-2. Problem (2-2) is called the *dual linear programming problem*. The two problems are called duals of each other and either one might be designated as the primal problem, depending upon which one of the problems is first formulated. It can be seen that problem (2-2) possesses all the characteristics of a linear programming problem and as such will yield an optimal solution if such exists.

Although many details will be given later, it should be noticed for the present that for this case the primal problem is a *maximization problem with equal-to-or-less-than constraints*, while the dual problem is a *minimization problem with equal-to-or-greater-than constraints*. Furthermore, the number of constraints in the primal problem is equal to the number of variables in the dual problem, and vice versa. Additional factors to be noted are that the successive coefficients of x_1 in each of the three constraints of the primal problem become the coefficients of y_1, y_2, and y_3 in the first constraint of the dual problem. The right-hand sides of the constraints in the primal problem are the coefficients of the objective function in the dual problem, and vice versa.

A number of very important relationships connect these two problems. These will be established once we build the proper background. *A very fundamental property is that, at optimality, the values of the two objective functions are equal.* Once we have obtained the solutions to these two problems, it will become evident that whenever we solve either of the problems, we also have the complete solution to the other problem. It remains, then, to be shown just why we are ever interested in the dual problem if we have the solution to the primal problem. Generally speaking, the two problems are not solved with the same degree of ease, and therefore there is a preference, quite marked in some cases, as to which form of the problem will be used for the solution. Criteria will be developed later for making this choice.

If one solves the primal problem stated in (2-1), it should be apparent that the entire decision-making process depends upon the following quantitative elements obtained from Table 2-1 on p. 14.

1. The rectangular array in the body of the table, which is outlined in heavy black lines. This portion of the table is in hours per piece.
2. The right-hand column of three figures, which gives the hourly restrictions for each of the machine centers.
3. The bottom row of four figures, which gives the per unit profit for each of the four products.

The array in element 1 will from now on be referred to as a matrix with three rows and four columns, and will have a designating label, such as

$$\mathbf{A} = \begin{bmatrix} 1.5 & 1 & 2.4 & 1 \\ 1 & 5 & 1 & 3.5 \\ 1.5 & 3 & 3.5 & 1 \end{bmatrix}$$

The column of figures in element 2 will be referred to as a column vector and may be labeled as

$$\mathbf{b} = \begin{bmatrix} 2000 \\ 8000 \\ 5000 \end{bmatrix}$$

The row of figures in element 3 will be referred to as a row vector and may be labeled

$$\mathbf{c} = (5.24, 7.30, 8.34, 4.18)$$

We must now provide precise definitions of matrix and vector as used above and show exactly how they can be used to advantage in working with linear programming problems.

2-4 MATRICES

In the material that follows, no attempt will be made to be completely rigorous. The topics from linear or matrix algebra that will be needed for later discussions will be presented briefly but in a manner in which they will be useful.

Matrix A *matrix* is defined to be a rectangular array of numbers and is written

$$\mathbf{A} = \|a_{ij}\| = \begin{bmatrix} a_{11} & a_{12} & a_{13} & \cdots & a_{1n} \\ a_{21} & a_{22} & \cdots & \cdots & a_{2n} \\ \cdots & \cdots & \cdots & \cdots & \cdots \\ a_{m1} & a_{m2} & a_{m3} & \cdots & a_{mn} \end{bmatrix} \tag{2-3}$$

The array is referred to as an $m \times n$, or m by n matrix. Brackets [ ], ‖ ‖, or parentheses ( ) are used to enclose such rectangular arrays or matrices. A matrix *does not* have a numerical value; it is merely an array. The numbers in the array are the elements of the matrix and the double subscript is used to signify its location, the first subscript indicating the row and the second subscript indicating the column. Whenever the entire matrix is represented by a single element, the notation $\|a_{ij}\|$, as shown above, is used.

Matrices will most often be denoted by boldface uppercase letter (**A**, **B**, **C**, etc.) and the individual elements will be denoted by lowercase letters (a_{ij}, b_{ij}, c_{ij}, etc.) unless the specific numbers themselves are used. Matrices may have the special form of consisting of either a single row or a single column and in such cases are referred to as *vectors*. If a matrix has the same number of rows as columns, it is called a *square matrix* and if it has n rows and n columns, it will be referred to as an nth-order square matrix.

We also define:

Equality of matrices Matrix **A** is *equal* to matrix **B** if and only if corresponding elements are equal; that is,

$$a_{ij} = b_{ij} \qquad \text{for all } i \text{ and } j$$

Addition of matrices The *sum* **C** of two matrices **A** and **B** is defined by the relationship

$$c_{ij} = a_{ij} + b_{ij} \qquad \text{for all } i \text{ and } j$$

From the definitions it is evident that two matrices can be equal only if they have the same size and shape, and the same must be true if two matrices are to be added. Furthermore, it should be evident that both the commutative and associative laws hold if the elements of the matrices are real numbers; that is,

$$\mathbf{A} + \mathbf{B} = \mathbf{B} + \mathbf{A}$$

$$\mathbf{A} + \mathbf{B} + \mathbf{C} = \mathbf{A} + (\mathbf{B} + \mathbf{C}) = (\mathbf{A} + \mathbf{B}) + \mathbf{C}$$

since addition does proceed by elements and these laws hold for elements that are real numbers.

Multiplication by a scalar The *product* of a matrix **A** by a scalar s is obtained by multiplying each element of **A** by s; that is,

$$s\mathbf{A} = \|sa_{ij}\|$$

2-5 EXAMPLES OF SIMPLE MATRIX OPERATIONS

Let

$$\mathbf{A} = \begin{bmatrix} 4 & 1 & -3 \\ 0 & -2 & 1 \end{bmatrix} \qquad \mathbf{B} = \begin{bmatrix} 2 & -1 & 7 \\ 4 & 2 & 3 \end{bmatrix} \qquad s = 0.3$$

Then by adding corresponding elements of the two matrices, we find that

$$\mathbf{A} + \mathbf{B} = \begin{bmatrix} 6 & 0 & 4 \\ 4 & 0 & 4 \end{bmatrix}$$

and by multiplying each of the elements of the matrix **B** by the scalar s, we obtain

$$s\mathbf{B} = \begin{bmatrix} 0.6 & -0.3 & 2.1 \\ 1.2 & 0.6 & 0.9 \end{bmatrix}$$

In order to present some logic to the definition of the product of two matrices, we find it convenient first to define the product of two vectors.

A vector, whether it be a row vector or a column vector, is a special case of a matrix; in the first case it has one row and in the second case it has one column. In either case it can be considered to be a point in n-dimensional space, where the components are the coordinates. Therefore the row vector (a_1, a_2, a_3) or the column vector

$$\begin{bmatrix} a_1 \\ a_2 \\ a_3 \end{bmatrix}$$

can be considered to be a point in three-dimensional space. There is no geometrical distinction between these two vectors. The decision to write a vector as a row vector or a column vector is therefore a matter of convenience.

The operation of addition of two vectors or multiplication by a scalar follows the rules for matrices. For these two operations it is helpful, however, to give the geometric interpretation, restricting our example to two dimensions. Consider the vector (a_1, a_2), which can be thought of as a point in two-dimensional space. We sometimes enhance this representation by drawing a line segment from the origin to the point (a_1, a_2) and then indicate direction by placing an arrowhead at the point (a_1, a_2). Multiplication by the scalar 1.5, for instance, would merely extend the length of the directed

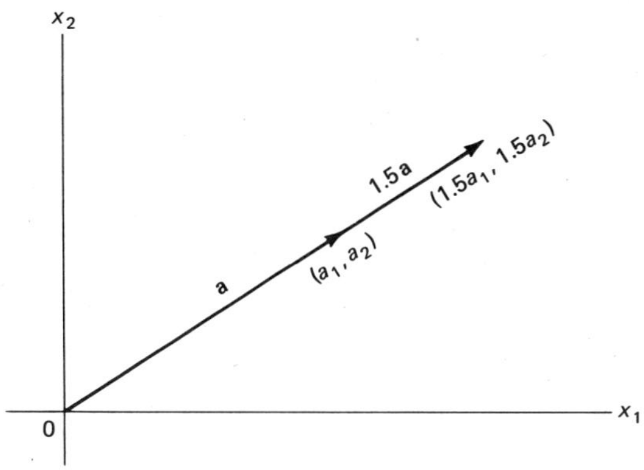

Figure 2-1

line segment in the same direction to the point that is 1.5 times as far from the origin. This is shown in Fig. 2-1.

Consider now the addition of two vectors **a** and **b**, where

$$\mathbf{a} = (a_1, a_2) \qquad \text{and} \qquad \mathbf{b} = (b_1, b_2)$$

Then

$$\mathbf{a} + \mathbf{b} = (a_1 + b_1, a_2 + b_2) = \mathbf{c}$$

Using the directed-line-segment approach, it can be seen from Fig. 2-2 that the sum **c** is the diagonal of the parallelogram having sides **a** and **b**. Similar graphical representations could be drawn in three-dimensional space, but higher dimensions require special treatment.

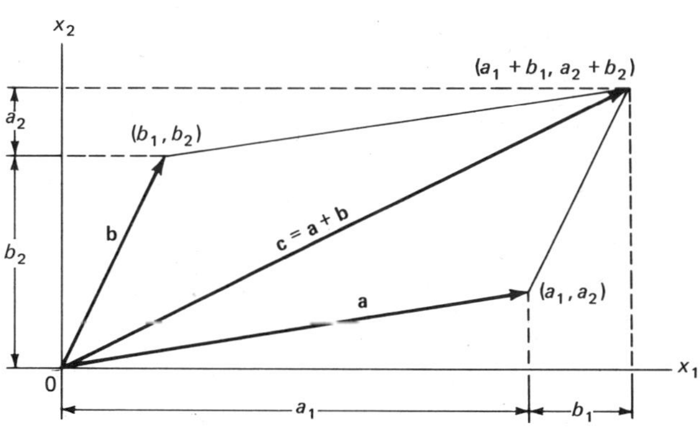

Figure 2-2

In the material that follows, we need two special vectors frequently. These are:

Unit vector A *unit vector*, denoted by $\mathbf{e}_i$, is a vector having its ith component equal to 1 and all other components equal to zero. Thus $\mathbf{e}_1 = (1, 0, 0)$, $\mathbf{e}_2 = (0, 1, 0)$, and $\mathbf{e}_3 = (0, 0, 1)$ are the unit row vectors for three-dimensional space.

Null vector The *null vector* is one that has each of its components equal to zero.

Inequalities are often used to indicate relationships between vectors. If $\mathbf{a} \leqq \mathbf{b}$, then $a_i \leqq b_i$ for every component i and similarly, if $\mathbf{a} \geqq \mathbf{b}$, then $a_i \geqq b_i$ for every component i.

2-6 SCALAR PRODUCT OF TWO VECTORS

The scalar product of two n-component vectors $\mathbf{a}$ and $\mathbf{b}$ is defined to be the single number given by

$$\sum_{i=1}^{n} a_i b_i$$

This is true whether $\mathbf{a}$ and $\mathbf{b}$ are row vectors or column vectors or even if we have one of each. If $\mathbf{a}$ is a row vector and $\mathbf{b}$ is a column vector, the product

$$\mathbf{ab} = (a_1, a_2, a_3)\begin{bmatrix} b_1 \\ b_2 \\ b_3 \end{bmatrix} = (a_1 b_1 + a_2 b_2 + a_3 b_3)$$

It is convenient for other reasons to insist that the row vector precede the column vector, and it is necessary that the number of components in each of the vectors be equal in order for the product to be defined.

The usefulness of this product can be shown in many ways. For example, suppose that $\mathbf{a} = (1.30, 4.20, 3.50)$ represents a selling-price vector, where the 1.30, 4.20, and 3.50 are the current selling prices for three items in the product line. Suppose further that

$$\mathbf{b} = \begin{bmatrix} 70 \\ 50 \\ 120 \end{bmatrix}$$

is a vector representing the amounts sold of each of these items. If we wish to obtain the total income from these items, we find the value of

$$\mathbf{ab} = (1.30, 4.20, 3.50)\begin{bmatrix} 70 \\ 50 \\ 120 \end{bmatrix}$$

which is

$$(1.30)(70) + (4.20)(50) + (3.50)(120) = 721$$

Thus it would be convenient to store this pertinent information in the form of a price vector and a sales-quantity vector to be held for easy evaluation of total sales income.

An extension of the information to many price vectors and to many sales-quantity vectors leads us to the need for multiplication of matrices. Using the same price vector as above but writing a number of sales-quantity vectors at different points in time by means of the matrix

$$\begin{bmatrix} 70 & 90 & 60 & 50 \\ 50 & 70 & 70 & 40 \\ 120 & 110 & 100 & 70 \end{bmatrix}$$

where each column of the matrix can be thought of as a sales-quantity vector at a different point in time, one could perform the multiplication of the price vector by each of the column vectors in turn, or in reality perform a vector-by-matrix multiplication, writing it as

$$(1.30, 4.20, 3.50)\begin{bmatrix} 70 & 90 & 60 & 50 \\ 50 & 70 & 70 & 40 \\ 120 & 110 & 100 & 70 \end{bmatrix} = (721, 796, 722, 478)$$

Then one could record each of the results as a row vector, as shown, so that the row vector represents the sales income at the points in time corresponding to those when the quantities were recorded. To extend the idea still further, we could consider the consequences of an additional price vector,

$$(1.40, 4.10, 3.60)$$

for example. To incorporate this price vector with the other price vector, we could write a price matrix

$$\begin{bmatrix} 1.30 & 4.20 & 3.50 \\ 1.40 & 4.10 & 3.60 \end{bmatrix}$$

by writing the second price vector underneath the first price vector. If we then wish to evaluate the sales income at the four points in time for each of the two price vectors, we would evaluate

$$\begin{bmatrix} 1.30 & 4.20 & 3.50 \\ 1.40 & 4.10 & 3.60 \end{bmatrix}\begin{bmatrix} 70 & 90 & 60 & 50 \\ 50 & 70 & 70 & 40 \\ 120 & 110 & 100 & 70 \end{bmatrix} = \begin{bmatrix} 721 & 796 & 722 & 478 \\ 735 & 809 & 731 & 486 \end{bmatrix}$$

which is really the product of two different price vectors by a quantity matrix consisting of three rows and four columns. By writing the second result underneath the first result, we have a matrix as a result. Thus we have actually carried out a matrix-by-matrix multiplication and need only to formalize the definition.

2-7 MULTIPLICATION OF MATRICES

We are now in position to define the product of two matrices **A** and **B** by making use of the elements of vector multiplication as just shown. The matrix product **AB** in that order is defined if and only if the number of columns of **A** is equal to the number of rows of **B**. There is no restriction on either the number of rows of **A** or the number of columns of **B**. If **A** is an m by r matrix and **B** is an r by n matrix, the product $\mathbf{C} = \mathbf{AB}$ is defined and **C** is an m by n matrix whose elements are computed from

$$c_{ij} = \sum_{k=1}^{r} a_{ik} b_{kj} \qquad i = 1, \ldots, m \text{ and } j = 1, \ldots, n$$

To illustrate again, consider the example

$$\mathbf{A} = \begin{bmatrix} 2 & 3 \\ 3 & 5 \\ 1 & 2 \end{bmatrix} \qquad \mathbf{B} = \begin{bmatrix} 2 & 3 & 0 & 1 \\ 1 & 0 & 2 & 3 \end{bmatrix}$$

For the product **AB** the element in the first row and the first column of the result is the product of the row vector (2, 3) from **A** and the column vector

$$\begin{bmatrix} 2 \\ 1 \end{bmatrix}$$

from **B**; that is,

$$(2, 3)\begin{bmatrix} 2 \\ 1 \end{bmatrix} = 7$$

The row vector (2, 3) is also used to obtain the remainder of the elements in the first row of the resulting product matrix by applying it to the remaining column vectors,

$$\begin{bmatrix} 3 \\ 0 \end{bmatrix} \qquad \begin{bmatrix} 0 \\ 2 \end{bmatrix} \qquad \begin{bmatrix} 1 \\ 3 \end{bmatrix}$$

successively. The elements of the second row of the product matrix are obtained similarly by using the second row of the matrix **A** as the preceding row vector and the same columns from **B** as the column vectors. This process is repeated for each row of the preceding matrix. It can be concluded that the number of rows in the resulting product matrix is equal to the number of rows in the preceding matrix and the number of columns in the resulting product matrix is equal to the number of columns in the postmatrix. For example,

$$\begin{bmatrix} 2 & 3 \\ 3 & 5 \\ 1 & 2 \end{bmatrix} \begin{bmatrix} 2 & 3 & 0 & 1 \\ 1 & 0 & 2 & 3 \end{bmatrix} = \begin{bmatrix} 7 & 6 & 6 & 11 \\ 11 & 9 & 10 & 18 \\ 4 & 3 & 4 & 7 \end{bmatrix}$$

Note that the product **BA** is not defined in this case, since the number of columns of **B** is not equal to the number of rows of **A**, and therefore the individual vector products do not exist. Thus we see that, in general, matrix multiplication does not satisfy the commutative law. It does, however, satisfy both the associative and distributive laws. The *associative* law for multiplication requires that identical results be obtained for the product **ABC** for the two orders

$$\mathbf{ABC} = (\mathbf{AB})\mathbf{C} = \mathbf{A}(\mathbf{BC})$$

which means that if we wish to find the product of three matrices, we can either multiply the first two matrices and then postmultiply it by the third matrix, or we can multiply the second and third matrices and then premultiply it by the first matrix. If the triple product actually exists, the corresponding matrices will be compatible in either order.

The *distributive* law for multiplication with respect to addition requires that identical results be obtained whether you add the products **AB** and **AC** or multiply **A** by the sum of **B** and **C**:

$$\mathbf{A}(\mathbf{B} + \mathbf{C}) = \mathbf{AB} + \mathbf{AC}$$

The preceding laws have been stated without proof, even though the proofs are essentially routine.

2-8 LINEAR PROGRAMMING IN MATRIX NOTATION

With this background we are now in position to restate problem (2-1) in matrix notation. In order to do this, let

$$\mathbf{A} = \begin{bmatrix} 1.5 & 1 & 2.4 & 1 \\ 1 & 5 & 1 & 3.5 \\ 1.5 & 3 & 3.5 & 1 \end{bmatrix}$$

be the matrix which shows the hourly requirements for each product in each machine center.

Let

$$\mathbf{x} = \begin{bmatrix} x_1 \\ x_2 \\ x_3 \\ x_4 \end{bmatrix} \quad \text{be the vector of solutions}$$

$$\mathbf{b} = \begin{bmatrix} 2000 \\ 8000 \\ 5000 \end{bmatrix} \quad \begin{matrix} \text{be the vector of time requirements} \\ \text{in hours for the scheduling period} \end{matrix}$$

$$\mathbf{c} = \quad (5.24, 7.30, 8.34, 4.18) \quad \text{be the profits vector}$$

Now we can state the problem in the notation of matrix algebra. It is:

Find

$$\mathbf{x} \geqq \mathbf{0}$$

such that

$$\mathbf{Ax} \leqq \mathbf{b} \tag{2-4}$$

and such that

$$z = \mathbf{cx} \text{ is a } \textit{maximum}$$

Let us check out statement (2-4) to make sure that every facet of the problem has been stated.

$$\mathbf{x} \geqq \mathbf{0} \text{ is equivalent to } \begin{bmatrix} x_1 \\ x_2 \\ x_3 \\ x_4 \end{bmatrix} \geqq \mathbf{0}$$

indicating that each component of the solution must be nonnegative.

$\mathbf{Ax} \leqq \mathbf{b}$ means

$$\begin{bmatrix} 1.5 & 1 & 2.4 & 1 \\ 1 & 5 & 1 & 3.5 \\ 1.5 & 3 & 3.5 & 1 \end{bmatrix} \begin{bmatrix} x_1 \\ x_2 \\ x_3 \\ x_4 \end{bmatrix} \leqq \begin{bmatrix} 2000 \\ 8000 \\ 5000 \end{bmatrix}$$

which when multiplied out translates into

$$\begin{aligned} 1.5x_1 + x_2 + 2.4x_3 + x_4 &\leqq 2000 \\ x_1 + 5x_2 + x_3 + 3.5x_4 &\leqq 8000 \\ 1.5x_1 + 3x_2 + 3.5x_3 + x_4 &\leqq 5000 \end{aligned}$$

the constraints of the problem.

$z = \mathbf{cx}$ means

$$(5.24, 7.30, 8.34, 4.18)\begin{bmatrix} x_1 \\ x_2 \\ x_3 \\ x_4 \end{bmatrix}$$

which translates into

$$z = 5.24x_1 + 7.30x_2 + 8.34x_3 + 4.18x_4$$

and is the objective function to be maximized. So it is clearly seen that matrix statement (2-4) is indeed a concise statement of problem (2-1).

From the above it is apparent that matrix statement (2-4) actually may represent any linear programming problem if the problem is in the form of a *maximization* with all *equal-to-or-less-than* constraints. Since it is always possible to do just that, statement (2-4) does, in fact, represent the general linear programming problem. To further clarify this concept, consider the mathematical model for Example 1-3. This example is a minimization problem, but all the constraints are not written as $\leqq$ constraints, and therefore we must make some changes.

The fourth constraint is

$$x_1 + 5x_2 \geqq 5$$

We can convert this to an $\leqq$ constraint by multiplying through by a *negative one* (-1), since this will reverse the sense of the inequality. The result for this case is

$$-x_1 - 5x_2 \leqq -5$$

By a similar operation, the fifth constraint becomes

$$-5x_1 - 2x_2 \leqq -10$$

Reversing the signs of our objection function the problem now reads:

Find

$$x_1, x_2 \geqq 0$$

such that

$$\begin{aligned} x_1 \qquad\quad &\leqq 3 \\ x_2 &\leqq 4 \\ x_1 + x_2 &\leqq 5 \\ -x_1 - 5x_2 &\leqq -5 \\ -5x_1 - 2x_2 &\leqq -10 \end{aligned}$$

and such that

$$z = -3x_1 - 2x_2 \text{ is a } \textit{maximum}$$

If we now define

$$\mathbf{x} = \begin{bmatrix} x_1 \\ x_2 \end{bmatrix} \qquad \mathbf{A} = \begin{bmatrix} 1 & 0 \\ 0 & 1 \\ 1 & 1 \\ -1 & -5 \\ -5 & -2 \end{bmatrix} \qquad \mathbf{b} = \begin{bmatrix} 3 \\ 4 \\ 5 \\ -5 \\ -10 \end{bmatrix} \qquad \mathbf{c} = (-3, -2)$$

From these definitions it should be clear that the problem can be written in matrix form as follows:

Find

$$\mathbf{x} \geqq \mathbf{0}$$

such that

$$\mathbf{Ax} \leqq \mathbf{b}$$

and such that

$$z = \mathbf{cx} \text{ is a } maximum$$

These two examples begin to show some of the elegance of a matrix representation. This is only the beginning, however. The first example had four original variables and three constraints, and the second example had two original variables and five constraints, yet the matrix statements of the two problems were exactly the same once the definitions of the vectors and matrices had been made.

To take advantage of the elegance and sophistication, a few additional ideas must be explored.

2-9 IDENTITY, NULL, AND TRANSPOSED MATRICES

Identity matrix An *identity matrix* is a matrix with 1s along the diagonal running from the upper left-hand corner to the lower right-hand corner and 0s elsewhere. It is a square matrix. It is designated by **I** or $\mathbf{I}_n$, where the subscript indicates the size of the matrix. The n is usually not necessary, since the size can be seen from the context in which it is used.

For example,

$$\mathbf{I} = \mathbf{I}_4 = \begin{bmatrix} 1 & 0 & 0 & 0 \\ 0 & 1 & 0 & 0 \\ 0 & 0 & 1 & 0 \\ 0 & 0 & 0 & 1 \end{bmatrix}$$

Direct multiplication shows that if **A** is an m by n matrix,

$$\mathbf{I}_m\mathbf{A} = \mathbf{A}\mathbf{I}_n = \mathbf{A}$$

The fact that a matrix is unaltered by either premultiplying or postmultiplying by a compatible identity matrix is the reason for its being called an "identity" matrix. It is analogous to the number 1 as used in the ordinary multiplication of real numbers. As an illustration, consider

$$\begin{bmatrix} 1 & 0 & 0 \\ 0 & 1 & 0 \\ 0 & 0 & 1 \end{bmatrix} \begin{bmatrix} 4 & 1 & 2 & 4 \\ 1 & -4 & -2 & 5 \\ 4 & 2 & 5 & -1 \end{bmatrix} = \begin{bmatrix} 4 & 1 & 2 & 4 \\ 1 & -4 & -2 & 5 \\ 4 & 2 & 5 & -1 \end{bmatrix}$$

and
$$\begin{bmatrix} 4 & 1 & 2 & 4 \\ 1 & -4 & -2 & 5 \\ 4 & 2 & 5 & -1 \end{bmatrix} \begin{bmatrix} 1 & 0 & 0 & 0 \\ 0 & 1 & 0 & 0 \\ 0 & 0 & 1 & 0 \\ 0 & 0 & 0 & 1 \end{bmatrix} = \begin{bmatrix} 4 & 1 & 2 & 4 \\ 1 & -4 & -2 & 5 \\ 4 & 2 & 5 & -1 \end{bmatrix}$$

As seen, each case produces the original matrix.

Null matrix A *null matrix* is a matrix with each of the elements equal to zero.

In matrix operations, we find that the null matrix acts very much like the number zero in the real number system; adding it to a matrix does not alter the matrix, and whenever a matrix is subtracted from itself, the null matrix results. Furthermore, multiplying by the null matrix results in a null matrix, but one must be *very careful* to note that the fact that the product of two matrices **A** and **B** results in the null matrix does not imply that either of the matrices is a null matrix. This may seem unusual but is only because we are used to drawing conclusions from similar algebraic statements. You should prove this to yourself by writing down two matrices, neither of which is a null matrix but for which the product is a null matrix. A little experimentation will enable you to do this. See Sec. 2-30 for an example.

Transposed matrix The *transpose* of a given matrix is that matrix obtained by interchanging rows and columns (i.e., the rows of the original matrix become the columns of the transposed matrix, and vice versa). The transpose will be denoted by $\mathbf{A}'$.

Without proof, it should be noted that the transpose of a product of two matrices is the product of the transposes of the matrices taken in the reverse order. That is,

$$(\mathbf{AB})' = \mathbf{B}'\mathbf{A}'$$

and directly from the definition, the transpose of a transpose is the original matrix itself:

$$(\mathbf{A}')' = \mathbf{A}$$

Further, it can be seen from the definition that the transpose of the sum of two matrices is the sum of the transposes of the two matrices:

$$(\mathbf{A} + \mathbf{B})' = \mathbf{A}' + \mathbf{B}'$$

Now that the transpose of a matrix has been defined and some of its properties have been indicated, we are in position to make a matrix statement of the *dual linear programming problem.* For the primal problem:

Find

$$\mathbf{x} \geqq \mathbf{0}$$

such that

$$\mathbf{Ax} \leqq \mathbf{b}$$

and such that

$$z = \mathbf{cx} \text{ is a } \textit{maximum}$$

Let us consider that there are m constraints and then define a column vector with m components, each component being associated with one of the constraints.

$$\mathbf{y} = \begin{bmatrix} y_1 \\ y_2 \\ y_3 \\ \vdots \\ y_m \end{bmatrix}$$

With this definition and our understanding of the transpose of a matrix, our dual statement of the general linear programming problem becomes:

Find

$$\mathbf{y} \geqq \mathbf{0}$$

such that

$$\mathbf{A}'\mathbf{y} \geqq \mathbf{c}' \tag{2-5}$$

and such that

$$Z = \mathbf{b}'\mathbf{y} \text{ is a } \textit{minimum}$$

If one now refers to the primal problem stated on p. 26, we have

$$\mathbf{y} = \begin{bmatrix} y_1 \\ y_2 \\ y_3 \end{bmatrix} \qquad \mathbf{A}' = \begin{bmatrix} 1.5 & 1 & 1.5 \\ 1 & 5 & 3 \\ 2.4 & 1 & 3.5 \\ 1 & 3.5 & 1 \end{bmatrix} \qquad \mathbf{c}' = \begin{bmatrix} 5.24 \\ 7.30 \\ 8.34 \\ 4.18 \end{bmatrix}$$

$$\mathbf{b}' = (2000, 8000, 5000)$$

and we find that $\mathbf{A}'\mathbf{y} \geqq \mathbf{c}'$ is actually

$$\begin{bmatrix} 1.5 & 1 & 1.5 \\ 1 & 5 & 3 \\ 2.4 & 1 & 3.5 \\ 1 & 3.5 & 1 \end{bmatrix} \begin{bmatrix} y_1 \\ y_2 \\ y_3 \end{bmatrix} \geqq \begin{bmatrix} 5.24 \\ 7.30 \\ 8.34 \\ 4.18 \end{bmatrix}$$

or

$$\begin{aligned} 1.5y_1 + \quad y_2 + 1.5y_3 &\geqq 5.24 \\ y_1 + \quad 5y_2 + \quad 3y_3 &\geqq 7.30 \\ 2.4y_1 + \quad y_2 + 3.5y_3 &\geqq 8.34 \\ y_1 + 3.5y_2 + \quad y_3 &\geqq 4.18 \end{aligned}$$

which are the constraints as stated in (2-2) on p. 17. The objective function is clearly

$$Z = (2000, 8000, 5000) \begin{bmatrix} y_1 \\ y_2 \\ y_3 \end{bmatrix}$$

which is then

$$Z = 2000y_1 + 8000y_2 + 5000y_3$$

At this point it is not much more than a formal exercise to write what we have designated as the dual linear programming problem, but later it will become extremely important, as indicated in the material on pp. 17 and 18. We must now proceed to some further definitions to prepare for the solution of linear programming problems by matrix methods.

2-10 MATRIX INVERSE

Matrix inverse Given a square matrix $\mathbf{A}$, if there exists a square matrix $\mathbf{B}$ which satisfies the relation $\mathbf{AB} = \mathbf{I}$, then $\mathbf{B}$ is the *inverse* of $\mathbf{A}$ and it is denoted by $\mathbf{A}^{-1}$. Furthermore, $\mathbf{BA} = \mathbf{I}$. Thus we have

$$\mathbf{AA}^{-1} = \mathbf{A}^{-1}\mathbf{A} = \mathbf{I}$$

To get some appreciation and understanding of the utility of the matrix inverse, consider the following linear system of equations. Let

$$\mathbf{A} = \begin{bmatrix} 1.5 & 1 & 2.4 \\ 1 & 5 & 1 \\ 1.5 & 3 & 3.5 \end{bmatrix} \qquad \mathbf{x} = \begin{bmatrix} x_1 \\ x_2 \\ x_3 \end{bmatrix} \qquad \mathbf{b} = \begin{bmatrix} 2000 \\ 8000 \\ 5000 \end{bmatrix} \tag{2-6}$$

Thus if we form

$$\mathbf{Ax} = \mathbf{b}$$

we really have the following system of three equations in three unknowns:

$$\begin{aligned} 1.5x_1 + x_2 + 2.4x_3 &= 2000 \\ x_1 + 5x_2 + x_3 &= 8000 \\ 1.5x_1 + 3x_2 + 3.5x_3 &= 5000 \end{aligned}$$

To obtain a solution to this system when written in the matrix form, we start with the equation

$$\mathbf{Ax} = \mathbf{b}$$

Multiplying both sides of the equation on the left by $\mathbf{A}^{-1}$, we obtain

$$\mathbf{A}^{-1}\mathbf{Ax} = \mathbf{A}^{-1}\mathbf{b}$$

Evaluating the left-hand side by using the associate law, we have

$$\mathbf{A}^{-1}\mathbf{Ax} = (\mathbf{A}^{-1}\mathbf{A})\mathbf{x} = \mathbf{Ix} = \mathbf{x}$$

and thus our solution is

$$\mathbf{x} = \mathbf{A}^{-1}\mathbf{b}$$

Thus, if we are able to find the inverse of the matrix of coefficients, it is easy to find the solution of the system of equations. Note that all matrices do not have inverses, yet whenever the inverse to a matrix can be found, it is unique. However, there exist several methods that can be used for finding an inverse. To develop a method of finding the inverse that is best suited for use in linear programming, we shall revert to a more fundamental way of solving a linear system of equations known as *Gaussian elimination*. For the system of equations given by (2-6), let us form a new matrix by using the matrix **A** and annexing to it the vector **b** using a dashed line as a means of visually separating the coefficients of the variables from the right-hand side.

$$\left[\begin{array}{ccc:c} 1.5 & 1 & 2.4 & 2000 \\ 1 & 5 & 1 & 8000 \\ 1.5 & 3 & 3.5 & 5000 \end{array}\right] \tag{2-7}$$

Even though we have no labels, we know full well that this can be thought of as representing three equations in three unknowns.

2-11 ALGEBRAIC SOLUTION OF A LINEAR SYSTEM OF EQUATIONS

The elements of the matrix (2-7) provide everything that is necessary for the solution of the system. Simple operations will be performed on the elements of this matrix, keeping in mind that we wish to maintain an equivalent system of three equations in three unknowns. The two operations that will be performed are: (1) dividing any row by a nonzero constant, and (2) adding any multiple of any row to any other row or subtracting any multiple of any row from any other row.

In the array above, if one divides the elements of the first row by 1.5, it then becomes

$$1 \qquad 0.66667 \qquad 1.60000 \qquad 1333.33$$

If this new first row is then subtracted from the original second row and if 1.5 times the new first row is subtracted from the original third row, the resulting matrix is

$$\left[\begin{array}{ccc|c} 1 & 0.66667 & 1.60000 & 1333.33 \\ 0 & 4.33333 & -0.60000 & 6666.67 \\ 0 & 2.00000 & 1.10000 & 3000.00 \end{array}\right] \tag{2-8}$$

This matrix still represents the original system of three equations in three unknowns, but it has been simplified in that the unknown x_1 is missing from the second and third equations.

We continue by eliminating x_2 from the first and third equations without destroying what we have done previously. To do this we divided the second row of (2-8) by 4.3333 and proceeded to reduce the second column to

$$\begin{bmatrix} 0 \\ 1 \\ 0 \end{bmatrix}$$

with the result

$$\left[\begin{array}{ccc|c} 1 & 0 & 1.69231 & 307.69 \\ 0 & 1 & -0.13846 & 1538.46 \\ 0 & 0 & 1.37692 & -76.92 \end{array}\right] \tag{2-9}$$

We continue by eliminating x_3 from the first and second equations, again without destroying what we have previously done. To do this we divide the third row by 1.37692 to make the element in the third row and third column unity. The other two elements in the third column are then made zeros by adding proper multiples of the new third row to each of the other rows, respectively, with the result

$$\left[\begin{array}{ccc|c} 1 & 0 & 0 & 402.23 \\ 0 & 1 & 0 & 1530.73 \\ 0 & 0 & 1 & -55.87 \end{array}\right] \tag{2-10}$$

We now have reduced the three equations in three unknowns to a form in which each equation has only one unknown. The solution can thus be read easily as

$$x_1 = 402.23$$
$$x_2 = 1530.73$$
$$x_3 = -55.87$$

By substitution, these values are seen to satisfy the original equations to within a possible rounding error.

Since all the decisions in the preceding process were determined by the elements in the first three columns, it would be possible to solve several systems of equations at the same time if the only differences among the systems were in the terms on the right-hand sides of the equations.

Suppose that we were interested in solving the system

$$\begin{aligned} 1.5x_1 + x_2 + 2.4x_3 &= 1000 \\ x_1 + 5x_2 + x_3 &= 5000 \\ 1.5x_1 + 3x_2 + 3.5x_3 &= 2000 \end{aligned}$$

and the system

$$\begin{aligned} 1.5x_1 + x_2 + 2.4x_3 &= 4000 \\ x_1 + 5x_2 + x_3 &= 1000 \\ 1.5x_1 + 3x_2 + 3.5x_3 &= 2000 \end{aligned}$$

while we are solving the system given by (2-7). We would then form

$$\left[\begin{array}{ccc:ccc} 1.5 & 1 & 2.4 & 2000 & 1000 & 4000 \\ 1 & 5 & 1 & 8000 & 5000 & 1000 \\ 1.5 & 3 & 3.5 & 5000 & 2000 & 2000 \end{array}\right]$$

and proceed to transform the first three columns into an identity matrix, obtaining

$$\left[\begin{array}{ccc:ccc} 1 & 0 & 0 & 402.23 & 1229.05 & 4435.75 \\ 0 & 1 & 0 & 1530.73 & 899.44 & -508.38 \\ 0 & 0 & 1 & -55.87 & -726.26 & -893.85 \end{array}\right]$$

from which we can read the solutions to each of three systems in columns four, five, and six, respectively. This process will be used in the following section to find the inverse of a matrix.

2-12 FINDING THE INVERSE OF A MATRIX

With the process described above in mind, we shall now develop one method of finding the inverse of a matrix. Consider the matrix **A** given in formulas (2-6) on p. 32. Let

$$\mathbf{A}^{-1} = \begin{bmatrix} a_{11} & a_{12} & a_{13} \\ a_{21} & a_{22} & a_{23} \\ a_{31} & a_{32} & a_{33} \end{bmatrix} \tag{2-11}$$

To find this inverse, it is necessary to find the a_{ij}'s that satisfy

$$\begin{bmatrix} 1.5 & 1 & 2.4 \\ 1 & 5 & 1 \\ 1.5 & 3 & 3.5 \end{bmatrix} \begin{bmatrix} a_{11} & a_{12} & a_{13} \\ a_{21} & a_{22} & a_{23} \\ a_{31} & a_{32} & a_{33} \end{bmatrix} = \begin{bmatrix} 1 & 0 & 0 \\ 0 & 1 & 0 \\ 0 & 0 & 1 \end{bmatrix} \tag{2-12}$$

and this is equivalent to the following three systems of equations:

$$\begin{bmatrix} 1.5 & 1 & 2.4 \\ 1 & 5 & 1 \\ 1.5 & 3 & 3.5 \end{bmatrix} \begin{bmatrix} a_{11} \\ a_{21} \\ a_{31} \end{bmatrix} = \begin{bmatrix} 1 \\ 0 \\ 0 \end{bmatrix} \tag{2-13}$$

and

$$\begin{bmatrix} 1.5 & 1 & 2.4 \\ 1 & 5 & 1 \\ 1.5 & 3 & 3.5 \end{bmatrix} \begin{bmatrix} a_{12} \\ a_{22} \\ a_{32} \end{bmatrix} = \begin{bmatrix} 0 \\ 1 \\ 0 \end{bmatrix} \tag{2-14}$$

and

$$\begin{bmatrix} 1.5 & 1 & 2.4 \\ 1 & 5 & 1 \\ 1.5 & 3 & 3.5 \end{bmatrix} \begin{bmatrix} a_{13} \\ a_{23} \\ a_{33} \end{bmatrix} = \begin{bmatrix} 0 \\ 0 \\ 1 \end{bmatrix} \tag{2-15}$$

The three systems have the same set of coefficients but different right-hand sides. Using the method discussed above, we form

$$\left[\begin{array}{ccc|ccc} 1.5 & 1 & 2.4 & 1 & 0 & 0 \\ 1 & 5 & 1 & 0 & 1 & 0 \\ 1.5 & 3 & 3.5 & 0 & 0 & 1 \end{array}\right] \tag{2-16}$$

and proceed with transformations that reduce the first three columns to an identity matrix. The result is

$$\left[\begin{array}{ccc|ccc} 1 & 0 & 0 & 1.6201 & 0.4134 & -1.2291 \\ 0 & 1 & 0 & -0.2235 & 0.1844 & 0.1006 \\ 0 & 0 & 1 & -0.5028 & -0.3352 & 0.7263 \end{array}\right] \tag{2-17}$$

Thus the inverse of the **A** matrix is found in the last three columns of (2-17). It is

$$\mathbf{A}^{-1} = \begin{bmatrix} 1.6201 & 0.4134 & -1.2291 \\ -0.2235 & 0.1844 & 0.1006 \\ -0.5028 & -0.3352 & 0.7263 \end{bmatrix} \tag{2-18}$$

It was shown on p. 32 that the solution to a linear system is given by

$$\mathbf{x} = \mathbf{A}^{-1}\mathbf{b}$$

and thus for the system (2-6),

$$\mathbf{x} = \mathbf{A}^{-1}\mathbf{b} = \begin{bmatrix} 1.6201 & 0.4134 & -1.2291 \\ -0.2235 & 0.1844 & 0.1006 \\ -0.5028 & -0.3352 & 0.7263 \end{bmatrix} \begin{bmatrix} 2000 \\ 8000 \\ 5000 \end{bmatrix} = \begin{bmatrix} 401.9 \\ 1531.2 \\ -55.7 \end{bmatrix} \qquad (2\text{-}19)$$

which is within a rounding error of the values found by the straightforward method on p. 34.

All matrices do not have inverses, but considerably more matrix algebra background is required to delineate the precise conditions under which the inverse does exist. It is sufficient to say that for the most part, we control the various aspects of a linear programming problem so that we do not have any difficulty due to lack of inverses.

2-13 EUCLIDEAN SPACE AND LINEAR INDEPENDENCE

The problems in linear programming with which we shall be concerned will be in what is called euclidean space, the kind of space with which we are most familiar. To give a formal definition of euclidean space, we need first to define distance between two points

$$\mathbf{a} = (a_1, a_2, \ldots, a_n) \qquad \text{and} \qquad \mathbf{b} = (b_1, b_2, \ldots, b_n)$$

It is

$$\text{distance} = \left[\sum_{i=1}^{n} (a_i - b_i)^2\right]^{1/2}$$

and is usually denoted by $|\mathbf{a} - \mathbf{b}|$. Then

Euclidean space An n-dimensional *euclidean space*, written E^n, is the collection of all n-dimensional vectors $\mathbf{a} = (a_1, a_2, \ldots, a_n)$ with the properties that the following exist and are defined:

1. Vector sum.
2. Multiplication of a vector by a scalar.
3. Distance as defined in the preceding paragraph.

The reader is undoubtedly familiar with our common two- and three-dimensional spaces of elementary geometry, which are euclidean spaces E^2 and E^3. We shall now develop some ideas which are at the very heart of linear programming.

Linear dependence A set of vectors $\mathbf{a}_1, \mathbf{a}_2, \ldots, \mathbf{a}_m$ selected from E^n is said to be *linearly dependent* if there exist scalars u_i not all zero such that

$$u_1 \mathbf{a}_1 + u_2 \mathbf{a}_2 + \cdots + u_m \mathbf{a}_m = 0$$

If $u_1 = u_2 = \cdots = u_m = 0$ is the only set for which the preceding equation is true, the vectors are said to be *linearly independent*. The terminology of linear dependence or independence is a property that is applied to a set of vectors and not to any one of the individual vectors of the set.

The idea of linear dependence as defined above is perfectly clear to the mathematically trained mind, but to others the definition perhaps seems rather cold and uninteresting, so we shall examine some of its properties and interesting features. As a result of the definition, we can show that a set of vectors from E^n will be linearly dependent if and only if one of the vectors can be written as a linear combination of the others. To illustrate the principle, consider the vectors

$$\begin{bmatrix} 3 \\ 2 \end{bmatrix} \quad \text{and} \quad \begin{bmatrix} 1 \\ 4 \end{bmatrix}$$

Are they linearly independent? To answer this question, we solve the following vector equation for the scalars u_1 and u_2:

$$u_1 \begin{bmatrix} 3 \\ 2 \end{bmatrix} + u_2 \begin{bmatrix} 1 \\ 4 \end{bmatrix} = \begin{bmatrix} 0 \\ 0 \end{bmatrix}$$

Simplifying, we have

$$\begin{bmatrix} 3u_1 \\ 2u_1 \end{bmatrix} + \begin{bmatrix} u_2 \\ 4u_2 \end{bmatrix} = \begin{bmatrix} 3u_1 + u_2 \\ 2u_1 + 4u_2 \end{bmatrix} = \begin{bmatrix} 0 \\ 0 \end{bmatrix}$$

and thus we must have

$$3u_1 + u_2 = 0$$

$$2u_1 + 4u_2 = 0$$

The solution to this system is $u_1 = u_2 = 0$, uniquely, and we conclude that

$$\begin{bmatrix} 3 \\ 2 \end{bmatrix} \quad \text{and} \quad \begin{bmatrix} 1 \\ 4 \end{bmatrix}$$

are linearly independent. From this you might say: "So what, isn't the same thing true for any two vectors?" Consider the vectors

$$\begin{bmatrix} 3 \\ 2 \end{bmatrix} \quad \text{and} \quad \begin{bmatrix} 6 \\ 4 \end{bmatrix}$$

Let us perform the same test. Solve for u_1 and u_2 in the equation

$$u_1 \begin{bmatrix} 3 \\ 2 \end{bmatrix} + u_2 \begin{bmatrix} 6 \\ 4 \end{bmatrix} = \begin{bmatrix} 0 \\ 0 \end{bmatrix}.$$

The solution is found in solving

$$3u_1 + 6u_2 = 0$$

$$2u_1 + 4u_2 = 0$$

Although $u_1 = u_2 = 0$ is a solution, it is not the only solution, for you can see that $u_1 = -2u_2$ in each of the equations, and therefore if we arbitrarily assign a value of 1 to u_2, we find that $u_1 = -2$. Thus for two dimensions we can see that if one of the vectors is a multiple of the other, the vectors are linearly dependent, so two vectors are *not* always linearly independent in E^2.

Consider now three vectors in two-dimensional space. Using the two vectors that we had before, which we know to be linearly independent, let us add a third vector and then test for linear independence. To make the test, we solve for u_1, u_2, and u_3 in

$$u_1 \begin{bmatrix} 3 \\ 2 \end{bmatrix} + u_2 \begin{bmatrix} 1 \\ 4 \end{bmatrix} + u_3 \begin{bmatrix} 2 \\ 2 \end{bmatrix} = \begin{bmatrix} 0 \\ 0 \end{bmatrix}$$

The solution is found by solving the linear system

$$3u_1 + u_2 + 2u_3 = 0$$

$$2u_1 + 4u_2 + 2u_3 = 0$$

A common method of solving this system is to write it in the form

$$3u_1 + u_2 = -2u_3$$

$$2u_1 + 4u_2 = -2u_3$$

In this form we can substitute an arbitrary value for u_3 and solve for u_1 and u_2. Thus if $u_3 = -5$, we find that $u_1 = 3$ and $u_2 = 1$. Since this is a nonzero solution, the set of vectors is linearly *dependent*. A little investigation will show that it is impossible to select a set of three vectors from two-dimensional space that are independent, since the method described above can always be used. In fact, it can be shown that whenever we select more than n vectors from an n-dimensional space, they must necessarily be linearly dependent. Let us now investigate some additional properties of linear independence.

2-14 SPECIAL PROPERTIES OF LINEAR INDEPENDENCE

It follows almost directly from the definition that if a given set of vectors is linearly independent, any subset of these vectors is also linearly independent.

It can also be shown quite easily that if a set of vectors is linearly dependent, any set including that set as a subset is also linearly dependent.

In the light of the previous discussion in reference to vectors in E^2, we noted that two vectors are linearly dependent if one is a scalar multiple of the other. In terms of our previous graphical representation, this means that the two points lie on the same straight line through the origin. Conversely, if any two vectors do not lie on the same straight line through the origin, they will be linearly independent.

In E^2, suppose we select two vectors that are linearly independent, such as

$$\begin{bmatrix} 3 \\ 2 \end{bmatrix} \quad \text{and} \quad \begin{bmatrix} 1 \\ 4 \end{bmatrix}$$

Then, if we select any other vector in E^2, it can be written as a linear combination of the two independent vectors. Suppose we select

$$\begin{bmatrix} 2 \\ 2 \end{bmatrix}$$

and write

$$\begin{bmatrix} 2 \\ 2 \end{bmatrix} = k_1 \begin{bmatrix} 3 \\ 2 \end{bmatrix} + k_2 \begin{bmatrix} 1 \\ 4 \end{bmatrix}$$

In order to find k_1 and k_2, we must solve the system

$$3k_1 + k_2 = 2$$
$$2k_1 + 4k_2 = 2$$

which yields

$$k_1 = \tfrac{3}{5}$$
$$k_2 = \tfrac{1}{5}$$

and therefore, we can write

$$\begin{bmatrix} 2 \\ 2 \end{bmatrix} = \tfrac{3}{5} \begin{bmatrix} 3 \\ 2 \end{bmatrix} + \tfrac{1}{5} \begin{bmatrix} 1 \\ 4 \end{bmatrix}$$

showing the selected vector written as a linear combination of the two linearly independent vectors. From this exercise we can see that any three vectors in two-dimensional space must necessarily be linearly dependent; for select two of the vectors, and they are either linearly dependent or linearly independent. If they are dependent, case proved, but if they are linearly independent, the third vector can be written as a linear combination of the other two, and hence the three vectors are linearly dependent.

If we extend these concepts to analogous ones in E^3, we find that any two vectors will lie in a plane through the origin in the three-dimensional space. Furthermore, any vector which is a linear combination of these two vectors, will lie in that same plane through the origin, and so if three vectors are linearly dependent, they will lie in a plane through the origin. If the three vectors are linearly independent, they will not lie in a plane through the origin. Any set of four vectors in E^3 can be shown to be linearly dependent.

The ideas presented above can be extended to higher spaces, but the interpretation is algebraic.

2-15 BASES AND SPANNING SETS OF VECTORS

The concept of a basis set of vectors is essential to a full understanding of linear programming. To understand this, we must define what is meant by a set of vectors that *span* the space.

> **Spanning set of vectors** A set of vectors $\mathbf{a}_1, \mathbf{a}_2, \ldots, \mathbf{a}_r$ selected from E^n will be said to *span* the space E^n if every vector in E^n can be written as a linear combination of these vectors.

It will now be necessary to determine the exact nature of a spanning set of vectors. With a little thought, it will become clear that *a complete set of unit vectors of any space* will span the space. To illustrate, consider E^3. The unit vectors are, of course, linearly independent, and therefore any arbitrarily selected vector can be written as a linear combination of these unit vectors; for example,

$$\begin{bmatrix} 1 \\ 2 \\ 4 \end{bmatrix} = 1 \begin{bmatrix} 1 \\ 0 \\ 0 \end{bmatrix} + 2 \begin{bmatrix} 0 \\ 1 \\ 0 \end{bmatrix} + 4 \begin{bmatrix} 0 \\ 0 \\ 1 \end{bmatrix}$$

Note that when the unit vectors are used as the spanning set, the scalar multiples of these unit vectors are the components of the vector itself. The spanning set of vectors is not unique, as we shall illustrate for the vector above. Suppose that we select the vectors

$$\begin{bmatrix} 1 \\ 2 \\ 3 \end{bmatrix}, \quad \begin{bmatrix} 4 \\ -1 \\ 2 \end{bmatrix}, \quad \text{and} \quad \begin{bmatrix} 3 \\ 1 \\ -2 \end{bmatrix}$$

These are seen to be linearly independent. Then, in writing

$$\begin{bmatrix}1\\2\\4\end{bmatrix} = k_1\begin{bmatrix}1\\2\\3\end{bmatrix} + k_2\begin{bmatrix}4\\-1\\2\end{bmatrix} + k_3\begin{bmatrix}3\\1\\-2\end{bmatrix}$$

we find that $k_1 = \frac{56}{49}$, $k_2 = \frac{5}{49}$, and $k_3 = -\frac{9}{49}$. Thus

$$\begin{bmatrix}1\\2\\3\end{bmatrix}, \quad \begin{bmatrix}4\\-1\\2\end{bmatrix}, \quad \text{and} \quad \begin{bmatrix}3\\1\\-2\end{bmatrix}$$

span the space, since we could have chosen any vector and found the proper linear combination. It is quite evident that *using the unit vectors* simplifies the determination of the linear combination.

The question then arises as to how many vectors are needed to span a space E^n. For an intuitive approach, consider the following logic. If we select an arbitrary vector from E^n, it has n components. How many vectors do we then need in the spanning set so that we may write the arbitrarily selected vector as a linear combination of that spanning set of vectors? If $\mathbf{a}_r$ is the arbitrarily selected vector and $\mathbf{a}_1, \ldots, \mathbf{a}_m$ are the vectors in the spanning set, we must write

$$\mathbf{a}_r = u_1\mathbf{a}_1 + u_2\mathbf{a}_2 + \cdots + u_m\mathbf{a}_m$$

and be able to solve for the u_i's uniquely. The foregoing vector equation results in n equations in m unknowns, $u_1, u_2, \ldots, u_m$. Using the intuitive approach, we know that if $m < n$, we will have more equations than we have unknowns, which we know are usually inconsistent. If $m > n$, we have more unknowns than we do equations, and the system usually has an infinite number of solutions. If $m = n$, we have a system with the same number of equations as we do unknowns, and we generally expect to be able to solve such a system uniquely. The exact criterion for obtaining a unique solution has to do with the rank of matrices, and we do not wish to get that theoretical here. Although that is not the only condition, your past experience indicates that $m = n$, meaning that it will be necessary to have exactly n vectors if you wish to obtain a unique solution for the set of u_i's.

A further question arises as to whether there are any requirements as to the nature of the vectors. Any spanning set of vectors that contains the least possible number of vectors will have to be linearly independent, since if the set were dependent, one of the vectors could be written as a linear combination of the others, and therefore it would not be required to form a spanning set. Since, as indicated on p. 38, the maximum number of linearly independent vectors in E^n is n, we shall need exactly n linearly independent vectors in the spanning set. We shall attach a special name to this set.

Basis A *basis* for an euclidean space E^n is a linearly independent set of vectors from E^n which spans the space.

It should be emphasized that a basis for a space E^n is *not* unique—in fact, there exist an infinite number of bases for any space. One of the sets that form a basis for the space is the set of unit vectors for that space. This was illustrated above for three-dimensional space. The set of unit vectors is commonly used as a basis for a space.

Even though the basis is not unique, *any vector in a space can be written in one and only one way as a linear combination of the given basis set*. This can be reasoned as follows. When writing the vector equation to find the linear combination of the basis set, one is led to a linear system of n equations in n unknowns. Although we do not prove it here, it can be shown that this solution is unique when the coefficients of the linear system come from the components of a set of linearly independent vectors.

In summary, we are led to the fact that if we wish to find a set of vectors from which we can generate all the vectors in a space E^n by forming a linear combination of the selected set, we must have exactly n vectors in the set, and the set must be linearly independent but is *not* unique. Each vector of the set is called a basic vector.

2-16 REPLACEMENT OF A BASIC VECTOR WITH AN ARBITRARY VECTOR

It will be brought out clearly that the technique of linear programming requires that for a given problem, considerable shifting be done from one set of basic vectors to another. In the process of this shifting, one of the most important and fundamental questions that must be answered is: *Under what conditions can an arbitrary vector* **b** *from E^n replace one of the vectors in a basis so that the new set of vectors also forms a basis?*

Consider an arbitrary vector **b** and a set of basis vectors $\mathbf{a}_1, \mathbf{a}_2, \ldots, \mathbf{a}_r$ from E^n. Since the set of vectors is independent, **b** can be written as a linear combination of the $\mathbf{a}_i$'s as follows:

$$\mathbf{b} = \sum_{i=1}^{r} m_i \mathbf{a}_i \tag{2-20}$$

If $\mathbf{b} \neq \mathbf{0}$, then at least one of the m_i's is nonzero. Suppose it to be m_r. We shall now show that the set $\mathbf{a}_1, \mathbf{a}_2, \ldots, \mathbf{a}_{r-1}, \mathbf{b}$ is also a basis for E^n; that is, the new set is linearly independent and it also spans the space. To prove independence, assume the contrary; that is, assume that

$$\sum_{i=1}^{r-1} \alpha_i \mathbf{a}_i + \alpha \mathbf{b} = 0 \tag{2-21}$$

and at least one α_i or α is nonzero. α cannot be zero, since this would contradict the fact that the set $\mathbf{a}_1, \mathbf{a}_2, \ldots, \mathbf{a}_{r-1}$ is linearly independent, since any subset of a linearly independent set of vectors is also linearly independent. By substituting (2-20) into (2-21), we obtain

$$\sum_{i=1}^{r-1} \alpha_i \mathbf{a}_i + \alpha \sum_{i=1}^{r} m_i \mathbf{a}_i = 0$$

and simplifying

$$\sum_{i=1}^{r-1} (\alpha_i + \alpha m_i)\mathbf{a}_i + \alpha m_r \mathbf{a}_r = 0 \tag{2-22}$$

in which $\alpha m_r \neq 0$, which then contradicts the preceding assumption that the set $\mathbf{a}_1, \mathbf{a}_2, \ldots, \mathbf{a}_r$ is linearly independent, and therefore the set $\mathbf{a}_1, \mathbf{a}_2, \ldots, \mathbf{a}_{r-1}, \mathbf{b}$ must be linearly independent.

If the set of vectors is to span the space, any arbitrarily selected vector $\mathbf{y}$ in E^n can be written as a linear combination of the set above. To show this, let

$$\mathbf{y} = \sum_{i=1}^{r} \beta_i \mathbf{a}_i \tag{2-23}$$

since the set $\mathbf{a}_1, \mathbf{a}_2, \ldots, \mathbf{a}_r$ forms a basis. From (2-20),

$$\mathbf{a}_r = \frac{1}{m_r}\left(\mathbf{b} - \sum_{i=1}^{r-1} m_i \mathbf{a}_i\right)$$

which, when substituted into (2-23), yields

$$\mathbf{y} = \sum_{i=1}^{r-1} \beta_i \mathbf{a}_i + \beta_r \left[\frac{1}{m_r}\left(\mathbf{b} - \sum_{i=1}^{r-1} m_i \mathbf{a}_i\right)\right]$$

Hence

$$\mathbf{y} = \sum_{i=1}^{r-1} \beta_i \mathbf{a}_i + \frac{\beta_r}{m_r}\mathbf{b} - \frac{\beta_r}{m_r}\sum_{i=1}^{r-1} m_i \mathbf{a}_i$$

and therefore

$$\mathbf{y} = \sum_{i=1}^{r-1} \left(\beta_i - \frac{\beta_r}{m_r} m_i\right)\mathbf{a}_i + \frac{\beta_r}{m_r}\mathbf{b} \tag{2-24}$$

in which $\mathbf{y}$ is expressed as a linear combination of the set $\mathbf{a}_1, \mathbf{a}_2, \ldots, \mathbf{a}_{r-1}, \mathbf{b}$, and therefore the set above forms a basis for E^n. It should be noted that if $\mathbf{b}$ is used to replace a vector $\mathbf{a}_i$ for which $m_i = 0$, the new set is linearly dependent and therefore does *not* form a basis. Thus if one is to select an arbitrary vector to replace one of a basis set of vectors, one must first write the selected vector as a linear combination of the basis set. The vector can then be used to replace any of the basis set of vectors that has a nonzero coefficient in this linear combination; otherwise, the new set will not be a basis set of vectors for that space.

2-17 NUMBER OF VECTORS IN A BASIS FOR E^n

As indicated in Sec. 2.15, every basis for E^n contains precisely n vectors. Some authors actually define the dimension of a space as the maximum number of linearly independent vectors in the space.

2-18 VECTOR SPACES AND SUBSPACES

From time to time we shall refer to vector spaces, so these should be defined.

Vector space A *vector space* is a collection of vectors such that the addition of any two of the vectors in the set yields a vector in the set, and also the multiplication of any vector in the set by a scalar yields a vector in the set.

Vector subspace A *subspace* of a vector space is a subset of the vector space which is itself a vector space.

2-19 RANK OF A MATRIX

Although we do not intend to prove many theorems dealing with the properties of matrices, we shall wish on occasion to make statements regarding certain properties for the more advanced reader. The concept of the rank of a matrix is often involved in some of these statements. For any m by n matrix, one can consider the n columns to be vectors in m-dimensional space. For convenience, then, these column vectors of $\mathbf{A}$ will be denoted by $\mathbf{a}_j$ and then

Rank of a matrix The *rank* of an m by n matrix $\mathbf{A}$, written $r(\mathbf{A})$, is the maximum number of linearly independent columns of $\mathbf{A}$. (This is also known as the *column rank* of the matrix.)

The *row rank* of the same matrix is defined as the maximum number of linearly independent rows in the matrix, but it can be shown that the row rank is equal to the column rank. Since the rank is unique, it is not necessary to distinguish between row rank and column rank.

In solving a system of linear equations, we are automatically interested in the rank of the matrix of coefficients, whether we realize it or not. If we have m equations in n unknowns, where $m \leq n$, the rank of the matrix of coefficients cannot possibly exceed m. If $m = n$ and the rank of the matrix is equal to m, the solution of the system is unique. If $m < n$, we know that there are an infinite number of solutions but the rank of the matrix may still be m.

This merely implies that we can select m column vectors which are linearly independent and that we are able to solve for the m corresponding variables uniquely in terms of the remaining variables. We shall find that we really do not need to test for rank in order to be able to carry out the appropriate steps in the solution of a linear programming problem. The algorithm that is used for the solution discovers whatever needs to be known about rank and handles it accordingly.

2-20 PRODUCT FORM OF THE INVERSE OF A MATRIX

In Sec. 2-12 a method was developed for obtaining the inverse of a square matrix. The application for this in linear programming problems is yet to come. As we solve more and more linear programming problems, we find that some of them become fairly complex and are also of considerable size. These larger problems are often solved by the revised simplex method. In using this method we have occasion to compute the inverse of a matrix for which only one column is different from that of a matrix whose inverse is already known. To develop the method for doing this, consider a matrix $\mathbf{B} = (\mathbf{b}_1, \mathbf{b}_2, \ldots, \mathbf{b}_n)$ whose inverse $\mathbf{B}^{-1}$ is already known. The $\mathbf{b}_i$'s above are column vectors. Suppose that we replace the rth column of $\mathbf{B}$ (that is, $\mathbf{b}_r$) by $\mathbf{a}$ and denote this new matrix by $\mathbf{B}_\mathbf{a} = (\mathbf{b}_1, \ldots, \mathbf{b}_{r-1}, \mathbf{a}, \mathbf{b}_{r+1}, \ldots, \mathbf{b}_n)$. We wish to find $\mathbf{B}_\mathbf{a}^{-1}$ from the information above.

Since the $\mathbf{b}_i$ represent a basis set of vectors, we can write $\mathbf{a}$ as a linear combination of the columns of $\mathbf{B}$:

$$\mathbf{a} = \sum_{i=1}^{n} y_i \mathbf{b}_i = \mathbf{B}\mathbf{y} \tag{2-25}$$

where

$$\mathbf{y} = \begin{bmatrix} y_1 \\ y_2 \\ \vdots \\ y_n \end{bmatrix}$$

Since $\mathbf{B}^{-1}$ exists, we can premultiply both sides of (2-25) by $\mathbf{B}^{-1}$ to obtain $\mathbf{B}^{-1}\mathbf{a} = \mathbf{B}^{-1}\mathbf{B}\mathbf{y} = \mathbf{I}\mathbf{y} = \mathbf{y}$. By expanding (2-25) and solving for $y_r \mathbf{b}_r$, we find that

$$y_r \mathbf{b}_r = -y_1 \mathbf{b}_1 \cdots - y_{r-1}\mathbf{b}_{r-1} + \mathbf{a} + \cdots - y_n \mathbf{b}_n$$

and if $y_r \neq 0$, then

$$\mathbf{b}_r = \frac{1}{y_r}(-y_1 \mathbf{b}_1 \cdots - y_{r-1}\mathbf{b}_{r-1} + \mathbf{a} - y_{r+1}\mathbf{b}_{r+1} \cdots - y_n \mathbf{b}_n)$$

Let us now define the column vector

$$\boldsymbol{\eta} = \begin{bmatrix} -\dfrac{y_1}{y_r} \\ \vdots \\ -\dfrac{y_{r-1}}{y_r} \\ \dfrac{1}{y_r} \\ -\dfrac{y_{r+1}}{y_r} \\ \vdots \\ -\dfrac{y_n}{y_r} \end{bmatrix}$$

With this definition, it can be seen that

$$\mathbf{b_r} = \mathbf{B_a}\boldsymbol{\eta}$$

It can also be seen that the following are true:

$$\mathbf{b}_1 = \mathbf{B_a}\mathbf{e}_1$$

$$\mathbf{b}_2 = \mathbf{B_a}\mathbf{e}_2$$

$$\cdots\cdots$$

$$\mathbf{b}_n = \mathbf{B_a}\mathbf{e}_n$$

where $\mathbf{e}_i$ is a column unit vector.

Therefore, if we define the matrix

$$\mathbf{E} = (\mathbf{e}_1, \ldots, \mathbf{e}_{r-1}, \boldsymbol{\eta}, \mathbf{e}_{r+1}, \ldots, \mathbf{e}_n)$$

we find that

$$\mathbf{B} = \mathbf{B_a}\mathbf{E}$$

Postmultiplying both sides by $\mathbf{B}^{-1}$ yields

$$\mathbf{B}\mathbf{B}^{-1} = \mathbf{I} = \mathbf{B_a}\mathbf{E}\mathbf{B}^{-1}$$

and now premultiplying both sides by $\mathbf{B_a}^{-1}$ results in

$$\mathbf{B_a}^{-1} = \mathbf{E}\mathbf{B}^{-1}$$

Now to summarize the procedure:

If the vector $\mathbf{a}$ is to replace the vector $\mathbf{b}_r$:

1. Compute $\mathbf{B}^{-1}\mathbf{a}$ to yield the vector $\mathbf{y}$.
2. Compute the vector $\boldsymbol{\eta}$ from the components of $\mathbf{y}$.
3. Define the matrix $\mathbf{E}$ using $\boldsymbol{\eta}$ and the appropriate unit vectors.
4. Compute $\mathbf{B_a}^{-1} = \mathbf{E}\mathbf{B}^{-1}$.

We can use this sequence of computations repeatedly to find the inverse of a matrix by replacing one column at a time, starting with an identity matrix whose inverse is also the same identity matrix.

2-21 ILLUSTRATIVE EXAMPLE OF AN INVERSE

Suppose we wish to invert the matrix **A** given in (2-6) by using this method:

$$\mathbf{A} = \begin{bmatrix} 1.5 & 1 & 2.4 \\ 1 & 5 & 1 \\ 1.5 & 3 & 3.5 \end{bmatrix}$$

We start with the identity matrix as **B**:

$$\mathbf{B} = \begin{bmatrix} 1 & 0 & 0 \\ 0 & 1 & 0 \\ 0 & 0 & 1 \end{bmatrix} \qquad \text{and} \qquad \mathbf{B}^{-1} = \mathbf{B}$$

Insert the first column of **A** in place of the first column of **B**. The first column of **A** is

$$\mathbf{a}_1 = \begin{bmatrix} 1.5 \\ 1 \\ 1.5 \end{bmatrix}$$

We first compute

$$\mathbf{y} = \mathbf{B}^{-1}\mathbf{a}_1 = \begin{bmatrix} 1 & 0 & 0 \\ 0 & 1 & 0 \\ 0 & 0 & 1 \end{bmatrix} \begin{bmatrix} 1.5 \\ 1 \\ 1.5 \end{bmatrix} = \begin{bmatrix} 1.5 \\ 1 \\ 1.5 \end{bmatrix}$$

From this we compute the vector $\boldsymbol{\eta}$ as follows.

Since we are replacing the first vector, the first position of $\boldsymbol{\eta}$ is the reciprocal of y_1 and all other components are the negative of y_i divided by y_1. The result is

$$\boldsymbol{\eta} = \begin{bmatrix} 1/1.5 \\ -1/1.5 \\ -1 \end{bmatrix}$$

The matrix **E** has the vector $\boldsymbol{\eta}$ in column one and unit vectors in each of the other columns. For this example

$$\mathbf{E} = \begin{bmatrix} 1/1.5 & 0 & 0 \\ -1/1.5 & 1 & 0 \\ -1 & 0 & 1 \end{bmatrix} = \begin{bmatrix} 0.6667 & 0 & 0 \\ -0.6667 & 1 & 0 \\ -1 & 0 & 1 \end{bmatrix}$$

$$\text{and} \quad \mathbf{B}_{\mathbf{a}}^{-1} = \mathbf{E}\mathbf{B}^{-1} = \begin{bmatrix} 0.6667 & 0 & 0 \\ -0.6667 & 1 & 0 \\ -1 & 0 & 1 \end{bmatrix} \begin{bmatrix} 1 & 0 & 0 \\ 0 & 1 & 0 \\ 0 & 0 & 1 \end{bmatrix} = \begin{bmatrix} 0.6667 & 0 & 0 \\ -0.6667 & 1 & 0 \\ -1 & 0 & 1 \end{bmatrix}$$

This becomes our new $\mathbf{B}^{-1}$, from which we now work. Having obtained an inverse with the first column replaced, we shall now replace the second column of this matrix with

$$\mathbf{a}_2 = \begin{bmatrix} 1 \\ 5 \\ 3 \end{bmatrix}.$$

The new $\mathbf{y}$ vector to be used for developing the new $\mathbf{E}$ matrix is now

$$\mathbf{y} = \mathbf{B}^{-1}\mathbf{a}_2 = \begin{bmatrix} 0.6667 & 0 & 0 \\ -0.6667 & 1 & 0 \\ -1 & 0 & 1 \end{bmatrix} \begin{bmatrix} 1 \\ 5 \\ 3 \end{bmatrix} = \begin{bmatrix} 0.6667 \\ 4.3333 \\ 2.0000 \end{bmatrix}$$

and
$$\mathbf{E} = \begin{bmatrix} 1 & -0.6667/4.3333 & 0 \\ 0 & 1/4.3333 & 0 \\ 0 & -2.0000/4.3333 & 1 \end{bmatrix} = \begin{bmatrix} 1 & -0.1538 & 0 \\ 0 & 0.2308 & 0 \\ 0 & -0.4615 & 1 \end{bmatrix}$$

and
$$\mathbf{B}_{\mathbf{a}}^{-1} = \begin{bmatrix} 1 & -0.1538 & 0 \\ 0 & 0.2308 & 0 \\ 0 & -0.4615 & 1 \end{bmatrix} \begin{bmatrix} 0.6667 & 0 & 0 \\ -0.6667 & 1 & 0 \\ -1.0000 & 0 & 1 \end{bmatrix}$$

$$= \begin{bmatrix} 0.7692 & -0.1538 & 0 \\ -0.1538 & 0.2308 & 0 \\ -0.6923 & -0.4615 & 1 \end{bmatrix}$$

The new $\mathbf{y}$ vector to be used for developing the third column of the next $\mathbf{E}$ is now

$$\mathbf{y} = \mathbf{B}^{-1}\mathbf{a}_3 = \begin{bmatrix} 0.7692 & -0.1538 & 0 \\ -0.1538 & 0.2308 & 0 \\ -0.6923 & -0.4615 & 1 \end{bmatrix} \begin{bmatrix} 2.4 \\ 1 \\ 3.5 \end{bmatrix} = \begin{bmatrix} 1.6923 \\ -0.1385 \\ 1.3769 \end{bmatrix}$$

and
$$\mathbf{E} = \begin{bmatrix} 1 & 0 & -1.6923/1.3769 \\ 0 & 1 & 0.1385/1.3769 \\ 0 & 0 & 1/1.3769 \end{bmatrix} = \begin{bmatrix} 1 & 0 & -1.2291 \\ 0 & 1 & 0.1006 \\ 0 & 0 & 0.7263 \end{bmatrix}$$

with the result that

$$\mathbf{B}_{\mathbf{a}}^{-1} = \begin{bmatrix} 1 & 0 & -1.2291 \\ 0 & 1 & 0.1006 \\ 0 & 0 & 0.7263 \end{bmatrix} \begin{bmatrix} 0.7692 & -0.1538 & 0 \\ -0.1538 & 0.2308 & 0 \\ -0.6923 & -0.4615 & 1 \end{bmatrix}$$

$$= \begin{bmatrix} 1.6201 & 0.4134 & -1.2291 \\ -0.2235 & 0.1844 & 0.1006 \\ -0.5028 & -0.3352 & 0.7263 \end{bmatrix}$$

which agrees with the result shown and developed in another way on p. 35. This method of inverting a matrix is used in the revised simplex procedure, which will be discussed later.

2-22 LINEAR SYSTEMS OF EQUATIONS

Returning now to the problem of the solution of more general systems of simultaneous linear equations, we shall let the coefficients of the unknowns be represented by $\mathbf{A} = \|a_{ij}\|$. The vector of right-hand-side constants is

$$\mathbf{b} = \begin{bmatrix} b_1 \\ b_2 \\ \vdots \\ b_m \end{bmatrix}$$

and the solutions vector is

$$\mathbf{x} = \begin{bmatrix} x_1 \\ x_2 \\ \vdots \\ x_n \end{bmatrix}$$

Then the system of equations expressed by the matrix equations

$$\mathbf{Ax} = \mathbf{b}$$

represents m equations in n unknowns. We have indicated previously that, if $m = n$ and the rank of $\mathbf{A}$ is equal to n, a unique solution is given by

$$\mathbf{x} = \mathbf{A}^{-1}\mathbf{b}$$

since $\mathbf{A}^{-1}$ itself is unique.

The case for which $m > n$ is of no interest to us in linear programming. The case that is most important to us is when $m < n$. As indicated before, such a system will have an infinite number of solutions. The algorithm for solving linear programming problems develops a criterion for the selection of the solution to be used.

2-23 BASIC SOLUTIONS

As just indicated, for linear programming purposes we are interested in linear systems in which there are more unknowns than there are equations. The extra unknowns are either slack variables or artificial variables, defined later, used in converting inequalities to equations. Among the infinite number of solutions that exist, we are especially interested in those solutions with as many of the variables as possible equal to zero. For a system with m equations in n unknowns, we shall assume that the rank of the matrix $\mathbf{A}$ is equal to m, for this will always be the case for a properly modeled linear programming problem.

By selecting any m linearly independent columns of $\mathbf{A}$, we can assign arbitrary values to the $n - m$ variables not associated with the selected columns. Suppose that we assign the value zero to each of these $n - m$ variables and solve for the remainder of the variables uniquely. These remaining m variables will, in general, have values different from zero, although it is not necessary for this to be true. The m variables in the solution which can be different from zero are called *basic variables* and the resulting complete solution is called a *basic solution.*

Using the notation that has been adopted by others, the matrix of the m linearly independent columns chosen from **A** will be designated by **B** and the basic solution will be written

$$\mathbf{x_B} = \mathbf{B}^{-1}\mathbf{b}$$

where all the variables in **A** which are not in **B** are called *nonbasic variables* and are set equal to zero. There are obviously a number of different basic solutions and hence a number of different matrices designated by **B**. We do not attempt to develop a notation to make the distinction, since the context of the problem makes it very clear which matrix is being represented by **B** at any particular time.

2-24 NUMBER OF BASIC SOLUTIONS

How many basic solutions are possible in a system of m equations in n unknowns? This is dependent upon the number of ways in which one can select a combination of m columns or variables from the n columns that are available for selection. This number of combinations is given by

$$\frac{n!}{m!\,(n - m)!}$$

which represents the largest possible number of basic solutions. The number will be less than this if some of the combinations are not linearly independent.

To illustrate the concept of basic solutions, consider the following system:

$$x_1 + 2x_2 + 3x_3 = 6$$

$$2x_1 + x_2 + 4x_3 = 4$$

The maximum number of basic solutions is three. Selection of a combination of two variables from among the three present is the same as selecting the one that is to be left out (i.e., the one that will be arbitrarily set to zero). Thus, if $x_1 = 0$, the solution to the remaining system is $x_2 = \frac{12}{5}$, $x_3 = \frac{2}{5}$. If $x_2 = 0$, the solution to the remaining variables is $x_1 = -6$, $x_3 = 4$. If $x_3 = 0$, the solution to the remaining system is $x_1 = \frac{2}{3}$, $x_2 = \frac{8}{3}$. Thus all three of the possible basic solutions exist.

2-25 DEGENERACY AND FEASIBILITY

In solving for basic solutions, we sometimes find that the basic variables also have zero values. If this is the case, the solution is referred to as a *degenerate solution.*

We sometimes find that the solutions to the basic variables are all nonnegative. We refer to this type of solution as a *feasible solution.* For the most part, we shall *not* be interested in nonfeasible solutions.

2-26 POINT SETS

In the material that follows, we shall speak of sets of points or vectors in an E^n space. Most of the time, the sets will contain an infinite number of elements and will be described by some well-defined property. In E^2, for example, the set of points lying inside a circle of radius 2 with its center at the origin is the set of points (x_1, x_2) satisfying the relation

$$x_1^2 + x_2^2 < 4$$

For this set the circle itself serves as a set of boundary points. If we wish to include the boundary points as a part of the set, we write

$$x_1^2 + x_2^2 \leqq 4$$

Whenever the set includes the set of boundary points, we refer to it as a *closed set*, and if it does not include its boundary points, we call it an *open set.*

2-27 STRAIGHT LINES AND HYPERPLANES

In E^2 the relationship

$$z = c_1 x_1 + c_2 x_2$$

represents a straight line for given values of c_1 and c_2 and z. This could be written $z = \mathbf{cx}$, where

$$\mathbf{x} = \begin{bmatrix} x_1 \\ x_2 \end{bmatrix} \quad \text{and} \quad \mathbf{c} = (c_1, c_2).$$

In E^3

$$z = c_1 x_1 + c_2 x_2 + c_3 x_3 = \mathbf{cx}$$

represents a plane.

In E^n

$$z = c_1 x_1 + c_2 x_2 + \cdots + c_n x_n = \mathbf{cx}$$

is called an *hyperplane*. In linear programming the set of all $\mathbf{x} = (x_1, x_2, \ldots, x_n)$ which gives a value to the objective function $z = \mathbf{cx}$ for some vector $\mathbf{c}$ is also an hyperplane.

Any hyperplane $z = \mathbf{cx}$ divides E^n into three mutually exclusive and exhaustive sets. These are:

1. The set for which $\mathbf{cx} < z$.
2. The set for which $\mathbf{cx} = z$.
3. The set for which $\mathbf{cx} > z$.

The first and third of these are called *open half-spaces*. If we add the first two sets together, we have a *closed half-space* (i.e., the set of points such that $\mathbf{cx} \leqq z$).

2-28 CONVEX SETS

A set of points $\mathbf{x}$ is said to be *convex* if for any two points $\mathbf{x}_1$ and $\mathbf{x}_2$ in the set, the line segment joining these two points is also in the set. This means that for $\mathbf{x}_1$ and $\mathbf{x}_2$ in the set, every point

$$\mathbf{x} = \mathbf{x}_2 + (1 - \lambda)\mathbf{x}_1$$

must also be in the set for $0 \leqq \lambda \leqq 1$. The term "convex" gives one an intuitive feeling for the nature of the set. Its boundaries are either flat or straight or curved away from the set, and furthermore there are no holes in its interior. The shaded areas of Figs. 1-1 to 1-4 show clearly the nature of a convex set with the boundaries being straight lines.

Because of its importance, we specifically define:

Extreme point A point $\mathbf{x}$ is an *extreme point* of a convex set if and only if there do not exist other points $\mathbf{x}_1$ and $\mathbf{x}_2$ in the set with $\mathbf{x}_1 \neq \mathbf{x}_2$ such that

$$\mathbf{x} = \mathbf{x}_2 + (1 - \lambda)\mathbf{x}_1$$

where $0 < \lambda < 1$.

This mathematical definition of an extreme point with λ necessarily being between zero and 1 states clearly that an extreme point of a convex set cannot be between any two points of the set. Because of this it is clear that an extreme point must lie on the boundary of the convex set, but every boundary point is not an extreme point—so it is a special kind of boundary point. For example, in two-dimensional space, a triangle and its interior form a convex set in which the sides are the boundary points but only the vertices are extreme points. Extreme points play an important part in the solution of linear programming problems.

We now have all the essentials for solving linear programming problems—matrices, vectors, and so on. Let us now look at a linear programming problem in the light of the newly established framework.

Each of the inequality constraints represents a closed half-space. The intersection of all these constraints, including the requirements for nonnegativity of the variables, results in a convex set of points for the feasible solution space. The objective function is a hyperplane which is moved over the convex set of points representing the feasible solution space so that at least one point in the hyperplane is always in the solution space.

2-29 LINEAR PROGRAMMING—CONCEPT OF A SOLUTION

If we now return to the linear programming problem formulated at the beginning of the chapter, one should be able to see that x_i, satisfying the three constraints, represents a closed half-space and each of the $x_i \geqq 0$ also represents a closed half-space. When all these constraints are applied (i.e., when the intersection of all these closed half-spaces is required) one should be able to see by analogy from the two-dimensional case that a convex set of points is obtained. We must be able to deal with this convex set of points algebraically.

To clarify our terminology we shall refer to any value of $\mathbf{x}$ which satisfies the linear constraints alone as a *solution*; if, in addition, it satisfies the nonnegativity requirements, it will be referred to as a *feasible solution.* In reference to problem statement (2-1), it is quite easy to select a set of values for x_1, x_2, x_3, and x_4 that will be a feasible solution; in fact, one might find a set of values with a substantial amount of profit. Without some kind of a test for optimality, however, one would always wonder if that is the best that can be done.

The three inequality constraints of the problem are not easy to handle in that form. To overcome the difficulties associated with inequalities, we shall introduce an additional variable in each of the inequality constraints in order to convert it into an equation. Thus with three new variables we now have three equations in seven variables as our constraints. The new variables introduced are real in the sense that a meaning can be assigned to each of them, each one of them representing the unused amount of the resource indicated by the right-hand side of the inequality.

In solving the linear programming problem we must always be able to find a solution to the three equations in seven unknowns thus obtained. The system obviously has an infinite number of solutions, but it only has a finite number of *basic* solutions. Basic solutions would be found by setting four of the variables equal to zero and solving for the remaining three, but there would be no guarantee that the solution would be feasible or optimal. It would be possible to find all the basic solutions, reject the nonfeasible ones, evaluate the objective function for each of those solutions retained, and then select the best from among these as the optimal basic feasible solution. This process would, of course, be prohibitive for a problem of any size. A better method must be developed.

As has already been mentioned, the set of feasible solutions to a linear programming problem is a closed convex set. The function to be optimized is a hyperplane. The graphical method of solution is that the hyperplane (representing the objective function) is moved over the convex set of solutions until the optimal value of z is found. During this movement the orientation of z is kept the same at all times, and at least one point of the solution set remains on the hyperplane. Thus at optimal, no point of the hyperplane could be an interior point of the set, and hence it must be a boundary point. Furthermore, the boundary point is an extreme point if there is a unique solution. If there are an infinite number of solutions, all of them are boundary points and at least one of these is an extreme point. Thus, if we could find a method of selecting the extreme points of a convex set of feasible solutions, we would have only a finite number of evaluations of the objective function to make.

The essential basis for the simplex method of solution developed in Chap. 3 is to find an initial extreme point which is basic and feasible. It is tested for optimality—if not optimal, we move to another extreme point, again testing for optimality and continuing the process until no further improvement can be made. Once the initial extreme point is selected, the movement to another extreme point is along the boundaries of the convex set of feasible solutions.

2-30 MATRIX EXAMPLE

On p. 29 it was indicated that it is possible to have the product of two matrices equal to zero without having either of the matrices equal to zero. To illustrate: If

$$\mathbf{A} = \begin{bmatrix} 2 & 4 \\ 1 & 2 \end{bmatrix} \quad \text{and} \quad \mathbf{B} = \begin{bmatrix} 2 & -4 \\ -1 & 2 \end{bmatrix}$$

we find that

$$\mathbf{AB} = \begin{bmatrix} 0 & 0 \\ 0 & 0 \end{bmatrix}$$

EXERCISES

The first three exercises are intended to give students practice in formulating linear programming problems. Each of these has a little different twist from the exercise introduced in the beginning of the chapter. The formulations should be saved for later solution.

1 *Product-mix problem.* A company has the technical knowledge and facilities to produce boxes and crates. The boxes will sell for \$3 each and the crates will sell for \$4 each. Each of the products must go through three operations in manufacturing: cutting, assembly, and testing.

A box requires:

2 min in cutting
2 min in assembly
1 min in testing

A crate requires:

3 min in cutting
1 min in assembly
1 min in testing

Each of these operations has a limited amount of time available during a scheduling period. The cutting time available is 140 min, the assembly time available is 80 min, and the testing time available is 50 min.

Write the mathematical statement of the problem which must be solved to find the product mix that will maximize the total sales dollars.

2 *Gasoline blending problem.* An oil company produces two grades of gasoline, which we shall call special and regular. The special sells for 69 cents/gal and the regular sells for 66 cents/gal. The refinery can buy two different crude oils containing ingredients M and N, which are critical in the blending of these two gasolines. The analysis of these two crudes, together with costs per gallon, are given in the table.

	Ingredients, %		
Oil	M	N	Cost/gal, cents
Crude 1	60	40	40
Crude 2	30	70	35

The regular gas must contain at least 55 percent of M. The special gas must contain at most 65 percent of N. The company's oil fields can provide up to 1,000,000 gal of crude 1 and 2,000,000 gal of crude 2 daily. The company says it can sell up to 800,000 gal of the regular gas and 3,500,000 gal of the special daily if it is available.

The nature of the problem might be thought of as follows. We would like to make a mixture of crude 1 and crude 2 to produce the special gas and another mixture of crude 1 and crude 2 to produce the regular gas. We shall not be concerned with just how the mixture is handled to produce the special and the regular gas, but some additional information might be helpful to clarify the previous table. Regardless of the actual process, suppose it is known that when crude 1 is directed to making special gas, the profit for this allocation is 29 cents/gal and if it is directed to the making of regular, the profit for this allocation is 26 cents/gal. The complete profit structure is shown in the table.

	Gasoline, cents		
	Special	Regular	Cost/gal, cents
Crude 1	29	26	40
Crude 2	34	31	35
Selling price	69	66	

Thus the profit on crude 1 directed to making regular gas is 66 cents less 40 cents, for a result of 26 cents. The others are found in a similar fashion. Formulate a linear programming problem that will maximize the daily profit from this operation.

3 *Shipping schedule problem.* A manufacturer makes a single product and has a monthly production capacity of 25,000 units. The unit production cost is \$8. The production must either be shipped or stored. If stored for later shipment, there is a storage cost of \$2/unit/month, the charge being imposed upon the quantity in storage at the end of the month.

Consider the following 4-month shipping requirement:

Month	Units
1	15,000
2	30,000
3	25,000
4	20,000

The company wishes to schedule production for the 4-month period so as to minimize the sum of the production and storage costs on the assumption that there was no beginning inventory and that the company wished to end the 4-month period with no inventory on hand.

Formulate a linear programming model that will minimize the sum of these costs.

The exercises that follow are intended to sharpen one's techniques in algebra and matrix algebra.

4 Given

$$\mathbf{A} = \begin{bmatrix} 1 & 2 & 3 \\ 4 & 2 & -1 \\ 1 & -1 & 2 \end{bmatrix} \qquad \mathbf{b} = \begin{bmatrix} 1 \\ -1 \\ 2 \end{bmatrix} \qquad \mathbf{a} = (1, 4, 5)$$

(*a*) Find **ab**.
(*b*) Find **Ab**.
(*c*) Find **aA**.
(*d*) Find $\mathbf{B} = \mathbf{A}^{-1}$.
(*e*) Find **AB** and show that it is the same as **BA**.

5 Solve the following system of linear equations using matrix methods:

$$\begin{aligned} x_1 + 2x_2 + 3x_3 &= 14 \\ 4x_1 + 2x_2 - x_3 &= 5 \\ x_1 - x_2 + 2x_3 &= 5 \end{aligned}$$

6 Write the mathematical statements of Exercises 1, 2, and 3 in matrix notation.

7 Test the following vectors for linear independence.

(*a*) $\begin{bmatrix} 3 \\ 2 \end{bmatrix}$ and $\begin{bmatrix} 2 \\ 4 \end{bmatrix}$.

(*b*) $\begin{bmatrix} 3 \\ 2 \end{bmatrix}$, $\begin{bmatrix} 2 \\ 4 \end{bmatrix}$ and $\begin{bmatrix} -1 \\ 4 \end{bmatrix}$.

(*c*) $\begin{bmatrix} 1 \\ 2 \\ 3 \end{bmatrix}$, $\begin{bmatrix} 4 \\ 2 \\ 1 \end{bmatrix}$, and $\begin{bmatrix} 1 \\ 1 \\ 1 \end{bmatrix}$

(*d*) $\begin{bmatrix} 1 \\ 2 \\ 3 \end{bmatrix}$, $\begin{bmatrix} 4 \\ 2 \\ 1 \end{bmatrix}$, $\begin{bmatrix} 1 \\ 1 \\ 1 \end{bmatrix}$, and $\begin{bmatrix} 2 \\ 4 \\ 2 \end{bmatrix}$.

(*e*) Show that any of the three vectors selected from part (*d*) are linearly independent. Write

$$\begin{bmatrix} 1 \\ 1 \\ 1 \end{bmatrix}$$

as a linear combination of the others.

8 (*a*) Find all the basic solutions of the following linear system of equations:

$$3x_1 + x_2 + 2x_3 = 4$$

$$2x_1 + 4x_2 + 4x_3 = 8$$

(*b*) Find all the basic *feasible* solutions to the system in part (*a*).

9 Find all the basic solutions to the following system of equations:

$$2x_1 + 4x_2 + 3x_3 + x_4 = 10$$

$$x_1 + 3x_2 + x_3 + 2x_4 = 15$$

$$8x_1 + 12x_2 + 5x_3 + 10x_4 = 35$$

In solving for the basic solutions, find those basic solutions in the following order:
Start with x_1, x_2, and x_3 as the basic variables.
Then in order replace x_3 with x_4.
Then replace x_2 with x_3.
Then replace x_1 with x_2.
Indicate which of the solutions are basic as well as feasible.
By selecting a different order of replacement, could you have found all the basic *feasible* solutions without finding any of the nonfeasible solutions?

10 Find the inverse of the matrix **A** in Exercise 4 using the product form of the inverse.

CHAPTER

THREE

SIMPLEX METHOD AND INTERPRETATION OF THE TABLEAUX

3-1 INTRODUCTION TO SIMPLEX METHOD—MATRIX STATEMENT

In the first two chapters we have given some indication of the kinds of problems we are to tackle under the linear programming framework. We then developed the basic background in linear algebra that would be needed for solving these problems efficiently. All linear programming problems deal with the optimization of a linear function subject to a set of linear constraints. The optimization may be either maximization or minimization. The linear constraints may be of the $\leqq$, $\geqq$, or $=$ variety. Rather than try to generalize right from the beginning, we shall first *concentrate on the solution of a problem of maximization subject to* $\leqq$ *constraints only*. Once we have handled this problem in detail, we shall extend the method to all other forms of the problem. In that way we do not introduce too many complications while trying to understand the basic method of solution.

In order to develop the basic ideas in the simplex method† of solution of linear programming problems, we shall return to the example stated at the beginning of Chap. 2 and use the example as a vehicle for introducing the

† The simplex method consists of moving from one basic feasible solution to another until optimality has been reached.

method, after which the principal ideas will be presented in more general terms. For our example, let

$$\mathbf{A} = \begin{bmatrix} 1.5 & 1 & 2.4 & 1 \\ 1 & 5 & 1 & 3.5 \\ 1.5 & 3 & 3.5 & 1 \end{bmatrix} \qquad \mathbf{x} = \begin{bmatrix} x_1 \\ x_2 \\ x_3 \\ x_4 \end{bmatrix} \qquad \mathbf{b} = \begin{bmatrix} 2000 \\ 8000 \\ 5000 \end{bmatrix}$$

$$\mathbf{c} = (5.24, 7.30, 8.34, 4.18)$$

The problem stated in matrix notation is then:

Find

$$\mathbf{x} \geqq \mathbf{0}$$

such that

$$\mathbf{Ax} \leqq \mathbf{b}$$

and

$$z = \mathbf{cx} \text{ is a } maximum$$

Specifically, the $\mathbf{Ax} \leqq \mathbf{b}$ is

$$\begin{bmatrix} 1.5 & 1 & 2.4 & 1 \\ 1 & 5 & 1 & 3.5 \\ 1.5 & 3 & 3.5 & 1 \end{bmatrix} \begin{bmatrix} x_1 \\ x_2 \\ x_3 \\ x_4 \end{bmatrix} \leqq \begin{bmatrix} 2000 \\ 8000 \\ 5000 \end{bmatrix}$$

and the $z = \mathbf{cx}$ is

$$(5.24, 7.30, 8.34, 4.18) \begin{bmatrix} x_1 \\ x_2 \\ x_3 \\ x_4 \end{bmatrix}$$

To apply the simplex method, the inequalities are first converted into equations by the addition of new variables. Since each of the inequalities is of the $\leqq$ variety, each of the new variables to be added is called *slack*, and for this problem, it represents the unused hours in each of the machine centers. We then have the equations:

$$\begin{aligned} 1.5x_1 + x_2 + 2.4x_3 + x_4 + x_5 \qquad\qquad &= 2000 \\ x_1 + 5x_2 + x_3 + 3.5x_4 \qquad + x_6 \qquad &= 8000 \\ 1.5x_1 + 3x_2 + 3.5x_3 + x_4 \qquad\qquad + x_7 &= 5000 \end{aligned} \tag{3-1}$$

where x_5, x_6, and x_7 represent the unused hours in machine centers *A*, *B*, and *C*, respectively. We shall make the assumption that unused hours contribute nothing to profit, and therefore the coefficients of these variables in the objective function will each be zero. Consequently, our new matrices will be as follows:

$$\mathbf{A} = \begin{bmatrix} 1.5 & 1 & 2.4 & 1 & 1 & 0 & 0 \\ 1 & 5 & 1 & 3.5 & 0 & 1 & 0 \\ 1.5 & 3 & 3.5 & 1 & 0 & 0 & 1 \end{bmatrix} \qquad \mathbf{x} = \begin{bmatrix} x_1 \\ x_2 \\ x_3 \\ x_4 \\ x_5 \\ x_6 \\ x_7 \end{bmatrix}$$

$$\mathbf{b} = \begin{bmatrix} 2000 \\ 8000 \\ 5000 \end{bmatrix} \qquad \text{and} \qquad \mathbf{c} = (5.24,\ 7.30,\ 8.34,\ 4.18,\ 0,\ 0,\ 0)$$

and an alternative statement of our problem with the preceding matrices is,

Find

$$\mathbf{x} \geqq \mathbf{0}$$

such that

$$\mathbf{Ax} = \mathbf{b}$$

and

$$z = \mathbf{cx} \text{ is a } maximum$$

3-2 BASIC FEASIBLE SOLUTIONS

We now have a set of three equations in seven unknowns and we are interested in the basic feasible solutions of this system. It can be shown that the extreme points of the convex set of feasible solutions are basic feasible solutions. *Hence the optimal solution will be contained in the set of basic feasible solutions,* of which there is a finite number. As a matter of fact, if we were to find all of these, it would require the solution of 7!/3!4! or 35 sets of three linear equations in three unknowns, the elimination of some of these as nonfeasible, and the evaluation of the objective function for each of the remainder. Such an examination of the basic solutions alone would not determine whether there is an unbounded solution or not.

As a partial illustration of this method, suppose that we select x_3, x_5, and x_6 as the basic variables. The resulting system of equations will be

$$\begin{aligned} 2.4x_3 + x_5 \qquad &= 2000 \\ x_3 \qquad + x_6 &= 8000 \\ 3.5x_3 \qquad\qquad &= 5000 \end{aligned}$$

The solution to this system is

$$x_3 = 1428.57$$

$$x_5 = -1428.57$$

$$x_6 = 6571.43$$

Since this is a nonfeasible basic solution, it will be given no further consideration and hence we do not even evaluate the objective function. If we were to continue with the method, 34 more systems of three equations in three unknowns would have to be solved and the objective function evaluated if the solution is feasible. To say the least, the method would be extremely inefficient, and because of this, it is not used. It would be highly desirable to be able to select the proper extreme point directly and thus make only one evaluation, but to date no way has been found to do this. Present solution methods are based upon finding a basic feasible solution and progressing to ever-improving basic feasible solutions, until the optimal is reached.

To clarify and develop the concepts, consider the matrix of the system of Eqs. (3-1) with the appropriate labels at the top of the tableau as follows:

x_1	x_2	x_3	x_4	x_5	x_6	x_7	b
1.5	1	2.4	1	1	0	0	2000
1	5	1	3.5	0	1	0	8000
1.5	3	3.5	1	0	0	1	5000

(3-2)

The matrix corresponding to the original variables x_1, x_2, x_3, and x_4 was initially designated as **A**. The matrix corresponding to the variables x_5, x_6, and x_7 is, of course, an identity matrix and the requirements vector is designated as **b**. With the constraints in equation form as in (3-2) the matrix corresponding to variables x_1 through x_7 will now be designated as **A**. The scalar x_1 is one of the components of the solution vector **x**. Associated with x_1 is a three-dimensional activity vector denoted $\mathbf{a}_1$, which is initially the column vector

$$\begin{bmatrix} 1.5 \\ 1 \\ 1.5 \end{bmatrix}.$$

Each of the columns in tableau (3-2) will be referred to in the vector context as $\mathbf{a}_j$, the subscript j denoting that it is associated with the solution component x_j. Since tableau (3-2) contains an identity matrix and $\mathbf{b} \geqq \mathbf{0}$, a basic feasible solution for the system of equations can be read very easily:

$$x_5 = 2000$$

$$x_6 = 8000$$

$$x_7 = 5000$$

$$x_1 = x_2 = x_3 = x_4 = 0$$

The identity matrix contains the key to reading the nonzero portion of the basic solution, with the variables corresponding to each of the other columns taking on the value of zero. As we move from one basic solution to another, some three columns of the 3 by 7 matrix of coefficients will always be a third-order identity matrix. Also, we shall control the operations so as to keep the components of the **b** vector nonnegative. The method of computation is set up to provide just that, and if such is not the case, an error has been made.

By the very definition of a basic solution, we know that for this problem, *no more than three* of the variables can be nonzero. We use the terminology that these are the three variables that are programmed for use. It is obvious that the profit is zero at this initial stage, since no one of the three variables programmed for use contributes anything to profit. The question that arises, then, is whether it is worthwhile to program any other variable for use and, if so, which one of the present variables—x_5, x_6, or x_7—must be removed from the program. A test must be devised to determine whether a new variable is to be programmed in place of one of the present variables, and the method of making the exchange must be developed.

3-3 NONBASIC VARIABLES AS A LINEAR COMBINATION OF BASIC VARIABLES

The key to both the test and the development is the fact that each of the nonbasic variables or corresponding activity vectors can be expressed as a linear combination of the basic variables or corresponding activity vectors.

The activity vectors corresponding to variables x_5, x_6, and x_7 are "unit" vectors and as such are linearly independent. Since they are linearly independent, they span the space, and hence every vector in E^3 can be written as a linear combination of these unit vectors. Consequently, the nonbasic vectors of this problem can be written as a linear combination of these three-dimensional unit vectors. For example, $\mathbf{a}_1$ can be written as a linear combination of the activity vectors $\mathbf{a}_5$, $\mathbf{a}_6$, and $\mathbf{a}_7$ as follows:

$$\mathbf{a}_1 = \begin{bmatrix} 1.5 \\ 1 \\ 1.5 \end{bmatrix} = 1.5\begin{bmatrix} 1 \\ 0 \\ 0 \end{bmatrix} + 1\begin{bmatrix} 0 \\ 1 \\ 0 \end{bmatrix} + 1.5\begin{bmatrix} 0 \\ 0 \\ 1 \end{bmatrix} = 1.5\mathbf{a}_5 + \mathbf{a}_6 + 1.5\mathbf{a}_7$$

all the information of which is obtained from the activity matrix **A** in tableau (3-2), which contains an identity matrix. This same information can be expressed in terms of the variables themselves:

$$x_1 = 1.5x_5 + x_6 + 1.5x_7$$

From this it can be seen that if a decision is made to program 1 unit of x_1, then 1.5 units of x_5 must be removed from the program along with 1 unit of

x_6 and 1.5 units of x_7 if the constraints are still to be satisfied. Since the profit attributed to 1 unit of x_1 is \$5.24, the net gain by programming 1 unit of x_1 will be its profit of \$5.24 less the amount of profit forfeited by removing 1.5 units of x_5, 1 unit of x_6, and 1.5 units of x_7, each of which contributes nothing to profit. This forfeited value can be expressed as the product of two vectors as follows:

$$(c_5, c_6, c_7)\begin{bmatrix}1.5\\1\\1.5\end{bmatrix} = (0, 0, 0)\begin{bmatrix}1.5\\1\\1.5\end{bmatrix} = \mathbf{c_B}\,\mathbf{a}_1$$

This scalar product will be denoted by z_1, the subscript 1 indicating that the column vector comes from column 1. The net gain by programming 1 unit of x_1 is therefore $c_1 - z_1$, which is equal to \$5.24. The vector $\mathbf{c_B}$ is used to denote the coefficients in the objective function for those terms which are currently in the basis. We could just as well have used $z_1 - c_1$ as an indicator of the profitability of entering x_1. In this case a negative sign indicates that it is profitable to enter the vector or program the variable. For reasons that will be indicated later, we shall use the form $z_j - c_j$ as the indicator for the jth variable. To maintain an orderly arrangement of the various items related to the problem, a tableau is made up in the following manner:

Initial tableau

Basis \ c_j		x_1 5.24	x_2 7.30	x_3 8.34	x_4 4.18	x_5 0.00	x_6 0.00	x_7 0.00	b 0000
$\mathbf{x_B}$	$\mathbf{c'_B}$								
x_5	0.00	1.5000	1.0000	2.4000	1.0000	1.0000	0.0000	0.0000	2000
x_6	0.00	1.0000	5.0000	1.0000	3.5000	0.0000	1.0000	0.0000	8000
x_7	0.00	1.5000	3.0000	3.5000	1.0000	0.0000	0.0000	1.0000	5000
$z_j - c_j$		−5.24	−7.30	−8.34	−4.18	0.00	0.00	0.00	0000

The profit vector for the variables in the basis (x_5, x_6, x_7) is listed over at the left under the label $\mathbf{c'_B}$, corresponding to the basis variables with the label $\mathbf{x_B}$. The row labeled c_j at the top of the tableau is the complete profit vector and will remain the same throughout the solution. The row labeled

$$z_j - c_j = \mathbf{c_B}\,\mathbf{a}_j - c_j$$

is our indicator of profitability. In this first stage, it will be noticed that the $(z_j - c_j)$'s are merely the negatives of the c_j's. Furthermore, it will be noted that $z_j = \mathbf{c_B}\, \mathbf{a}_j$, where $\mathbf{a}_j$ represents the jth activity vector, as shown in the initial tableau. The $\mathbf{a}_j$ vectors will be transformed from tableau to tableau as the basis vectors are changed. These activity vectors will be denoted by $\mathbf{y}_j$ in all tableaux after the initial one. The $\mathbf{y}_j$ for any tableau can be computed from the $\mathbf{a}_j$ in the initial tableau by multiplying on the left by the inverse of the current set of basis vectors (i.e., $\mathbf{y}_j = \mathbf{B}^{-1}\mathbf{a}_j$), showing that the $\mathbf{y}_j$'s change with the changes in $\mathbf{B}$. $\mathbf{B}$ is a square matrix in the initial tableau corresponding to the identity matrix in the current tableau.

3-4 SELECTION OF INCOMING VECTOR

An examination of the $z_j - c_j$ row in the initial tableau shows that a unit net profit of \$8.34 will be obtained by programming the variable x_3. The variable x_3 has been selected for programming since its unit net profit is greater than that for any other variable. The fact that x_3 has been selected for entry into the program is designated with an arrow, $\uparrow$. The $z_j - c_j$ does not indicate how many units of x_3 are to be programmed; consequently, further analysis is necessary to determine the number of units to be programmed as well as the total gain in profit that can be obtained by programming x_3 to the fullest extent possible while maintaining a feasible solution.

3-5 SELECTION OF OUTGOING VECTOR

The determination of the number of units to be programmed is dependent upon the vector

$$\begin{bmatrix} 2.4 \\ 1 \\ 3.5 \end{bmatrix},$$

which is the activity vector corresponding to x_3 and the present constraints vector

$$\begin{bmatrix} 2000 \\ 8000 \\ 5000 \end{bmatrix}.$$

At this stage it is easy to see that since 2000 h are available in machine center A and since each unit of x_3 requires 2.4 h in that center, 2000 divided by 2.4 or $833\frac{1}{3}$ is the maximum number of units of x_3 that can be processed through machine center A if that and no other product is to be made. Similarly, it can be seen that 8000 units can be processed through machine center B and

$1428\frac{4}{7}$ units through machine center C. Since the product must be processed through all three machine centers, the smallest of these three numbers dictates the maximum number of units that can be made—in this case, $833\frac{1}{3}$. We thus will program $833\frac{1}{3}$ units of product x_3. In order to reflect this in the tableau, the vector $\mathbf{a}_3$, which is presently

$$\begin{bmatrix} 2.4 \\ 1 \\ 3.5 \end{bmatrix}$$

must be transformed into

$$\begin{bmatrix} 1 \\ 0 \\ 0 \end{bmatrix}$$

the position of unity being dictated by the fact that 2.4 divided into the constraint component was the most restrictive of the three possibilities. We call 2.4 the pivot element† and designate it on the tableau by ( ). In order to transform the entire tableau, the operations that are performed are; first, divide the entire row by the number in the pivot position; second, selected multiples of the pivot row are added to other rows to transform the activity vector of the incoming variable into a unit vector with the 1 in the pivot position. The selected multiples may be positive or negative.

Continuing with our example, the first row of the tableau, excluding the c_j row, will be divided by 2.4; it then becomes

$$0.6250 \quad 0.4167 \quad 1.0000 \quad 0.4167 \quad 0.4167 \quad 0.0000 \quad 0.0000 \qquad 833.33$$

If this row is then subtracted from the original second row, the result is

$$0.3750 \quad 4.5833 \quad 0.0000 \quad 3.0833 \quad -0.4167 \quad 1.0000 \quad 0.0000 \qquad 7166.67$$

and if 3.5 times the row is subtracted from the original third row, the result is

$$-0.6875 \quad 1.5417 \quad 0.0000 \quad -0.4583 \quad -1.4583 \quad 0.0000 \quad 1.0000 \qquad 2083.33$$

With this portion of the computation completed, it is necessary to compute the value of the objective function and to test for possible further improvement of the solution by programming other variables. With the variable x_3 replacing the *outgoing* variable x_5, the basis variables are now x_3, x_6, and x_7. Therefore, the profit vector for these basis variables is

$$\mathbf{c_B} = (8.34, 0, 0) \qquad \text{and} \qquad z_j = \mathbf{c_B}\,\mathbf{y}_j$$

where $\mathbf{y}_j$ refers to the now transformed $\mathbf{a}_j$ as it appears in the second tableau.

† The pivot operation was carried out in Sec. 2.11 without specific reference to the pivot element.

The entry in the **b** vector column corresponding to $z_j - c_j$ is $\mathbf{c_B x_B} - z_0 = \mathbf{c_B x_B}$, since $z_0 = 0$, the initial value of the objective function. $\mathbf{x_B}$ refers to the current basic solution, and it can be computed by $\mathbf{x_B} = \mathbf{B}^{-1}\mathbf{b}$ with **B** and $\mathbf{B}^{-1}$ referring to the matrix of the basis set of vectors. At this stage

$$\mathbf{B} = \begin{array}{c} \begin{matrix} \mathbf{a}_3 & \mathbf{a}_6 & \mathbf{a}_7 \end{matrix} \\ \begin{bmatrix} 2.4 & 0 & 0 \\ 1 & 1 & 0 \\ 3.5 & 0 & 1 \end{bmatrix} \end{array} \qquad \mathbf{B}^{-1} = \begin{bmatrix} 0.4167 & 0.0000 & 0.0000 \\ -0.4167 & 1.0000 & 0.0000 \\ -1.4583 & 0.0000 & 1.0000 \end{bmatrix}$$

With this information tabulated in the same way as was the initial tableau, we have:

Second tableau

Basis \ c_j		x_1 5.24	x_2 7.30	x_3 8.34	x_4 4.18	x_5 0.00	x_6 0.00	x_7 0.00	b 0000
$\mathbf{x_B}$	$\mathbf{c'_B}$								
x_3	8.34	0.6250	0.4167	1.0000	0.4167	0.4167	0.0000	0.0000	833.33
x_6	0.00	0.3750	4.5833	0.0000	3.0833	−0.4167	1.0000	0.0000	7166.67
x_7	0.00	−0.6875	(1.5417)	0.0000	−0.4583	−1.4583	0.0000	1.0000	2083.33
$z_j - c_j$		−0.0275	−3.8250	0.0000	−0.7050	3.4750	0.0000	0.0000	6950
			↑					↓	

The insertion of x_3 into the program changed the solution values of x_6 and x_7, so the basic feasible solution is now

$$x_3 = 833.33$$

$$x_6 = 7166.67$$

$$x_7 = 2083.33$$

$$x_1 = x_2 = x_4 = x_5 = 0$$

The value of the objective function is \$6950, and it is displayed in the extreme lower right-hand corner of the tableau. The incoming and outgoing vectors are marked along with the pivot element for the next iteration. The outgoing vector is marked ↓.

3-6 SECOND ITERATION

Examination of the $z_j - c_j$ row indicates that x_2 is the next variable selected to be entered, since -3.8250 is the negative with the largest absolute value. Division of the components of the vector $\mathbf{y}_2$ into the corresponding elements of the constraint vector shows that the third row is the most constraining. Hence 1.5417 is the pivot element and the vector $\mathbf{y}_2$ must be transformed into

$$\begin{bmatrix} 0 \\ 0 \\ 1 \end{bmatrix}$$

x_7 will be the variable removed from the basis. To accomplish the change, the third row is divided by 1.5417. Then -0.4167 times the result is added to the first row and -4.5833 times that result is added to the second row. The $z_j - c_j$ row is again computed from $\mathbf{c_B}\,\mathbf{y}_j - c_j$, with the following results:

Third tableau

Basis	c_j	x_1 5.24	x_2 7.30	x_3 8.34	x_4 4.18	x_5 0.00	x_6 0.00	x_7 0.00	b 000
$\mathbf{x_B}$	$\mathbf{c'_B}$								
x_3	8.34	0.8108	0.0000	1.0000	0.5405	0.8108	0.0000	−0.2703	270.27
x_6	0.00	2.4189	0.0000	0.0000	(4.4459)	3.9189	1.0000	−2.9730	972.97
x_2	7.30	−0.4459	1.0000	0.0000	−0.2973	−0.9459	0.0000	0.6486	1351.35
$z_j - c_j$		−1.7332	0.0000	0.0000	−1.8422	−0.1432	0.0000	2.4811	12119
					↑		↓		

The principal part of the matrix is enclosed with heavy lines to offset it from the rest of the information, which is made up of labels and indicators. The c_j row at the top remains the same for each of the tableaux. The two columns at the left specify the basis variables and the $z_j - c_j$ row at the bottom is used to test for possible improvement in the solution. Any reference to either the vector $\mathbf{a}_j$ or its transformed $\mathbf{y}_j$, or the vector $\mathbf{b}$ or its transformed form, will refer to the appropriate three-component vector within the heavy lines. At a later point, the $z_j - c_j$ row will be included in an expanded form of the matrix. The solution now reads

$$x_3 = 270.27$$
$$x_6 = 972.27$$
$$x_2 = 1351.35$$
$$x_1 = x_4 = x_5 = x_7 = 0$$

and the profit is

$$\mathbf{c_B\, x_B} = (8.34,\ 0,\ 7.30)\begin{bmatrix} 270.27 \\ 972.97 \\ 1351.35 \end{bmatrix} = \$12119$$

3-7 THIRD ITERATION

Continuing, x_4 is the next variable to enter the basis, since $z_4 - c_4 = -1.8422$ is the negative with the largest absolute value. To determine the number of units of x_4 to be programmed, we ignore the -0.2973 in that column, since using that element as a pivot point would mean that we would be inserting a negative amount into the program, which would be contrary to one of the basic constraints for the problem. Using the other two components in that column, we find that $\mathbf{y}_4$ (the current activity vector corresponding to x_4) is to be transformed into

$$\begin{bmatrix} 0 \\ 1 \\ 0 \end{bmatrix},$$

which indicates that x_6 is the outgoing variable. After the transformation, the result is:

Fourth tableau

Basis \ c_j		x_1 5.24	x_2 7.30	x_3 8.34	x_4 4.18	x_5 0.00	x_6 0.00	x_7 0.00	b 000
$\mathbf{x_B}$	$\mathbf{c'_B}$								
x_3	8.34	(0.5167)	0.0000	1.0000	0.0000	0.3343	−0.1216	0.0912	151.98
x_4	4.18	0.5441	0.0000	0.0000	1.0000	0.8815	0.2249	−0.6687	218.84
x_2	7.30	−0.2842	1.0000	0.0000	0.0000	−0.6839	0.0669	0.4498	1,416.42
$z_j - c_j$		−0.7310	0.0000	0.0000	0.0000	1.4805	0.4143	1.2492	12522
		↑		↓					

At this stage the solution is

$$x_3 = 151.98$$
$$x_4 = 218.84$$
$$x_2 = 1416.42$$
$$x_1 = x_5 = x_6 = x_7 = 0$$

with a profit of \$12,522.

3-8 FOURTH ITERATION

Examination of the $z_j - c_j$ in the fourth tableau shows that further improvement can be made by inserting x_1 into the program and removing x_3. The resulting optimal tableau is:

Optimal tableau

Basis	c_j	x_1 5.24	x_2 7.30	x_3 8.34	x_4 4.18	x_5 0.00	x_6 0.00	x_7 0.00	b
$\mathbf{x_B}$	$\mathbf{c'_B}$								
x_1	5.24	1.0000	0.0000	1.9353	0.0000	0.6471	−0.2353	0.1765	294.11
x_4	4.18	0.0000	0.0000	−1.0529	1.0000	0.5294	0.3529	−0.7647	58.82
x_2	7.30	0.0000	1.0000	0.5500	0.0000	−0.5000	0.0000	0.5000	1500.00
$z_j - c_j$		0.0000	0.0000	1.4146	0.0000	1.9537	0.2424	1.3782	12,737

We have now reached the stage in which there are no negative terms in the $z_j - c_j$ row, which means that no variable presently not in the basis can be inserted so as to give improvement in the profit, and therefore the *solution is optimal.* The solution is

$$x_1 = 294.11$$
$$x_4 = 58.82$$
$$x_2 = 1500.00$$
$$x_3 = x_5 = x_6 = x_7 = 0$$

with a profit of \$12,737.

For this solution we note that there are exactly three zeros in the $z_j - c_j$ row. Since there are three basic variables, there are no alternate solutions—the solution is unique. If there should be a zero in the $z_j - c_j$ row

	c_j:	5.24	7.30	8.34	4.18	0.00	0.00	0.00	000
	Variables:	x_1	x_2	x_3	x_4	x_5	x_6	x_7	b
$\mathbf{x_B}$	$\mathbf{c'_B}$								
x_5	0.00	1.5000	1.0000	2.4000*	1.0000	1.0000	0.0000	0.0000	2,000.00
x_6	0.00	1.0000	5.0000	1.0000	3.5000	0.0000	1.0000	0.0000	8,000.00
x_7	0.00	1.5000	3.0000	3.5000	1.0000	0.0000	0.0000	1.0000	5,000.00
$z_j - c_j$		−5.2400	−7.3000	−8.3400	−4.1800	0.0000	0.0000	0.0000	0,000†
x_3	8.34	0.6250	0.4167	1.0000	0.4167	0.4167	0.0000	0.0000	833.33
x_6	0.00	0.3750	4.5833	0.0000	3.0833	−0.4167	1.0000	0.0000	7,166.67
x_7	0.00	−0.6875	1.5417*	0.0000	−0.4583	−1.4583	0.0000	1.0000	2,083.33
$z_j - c_j$		−0.0275	−3.8250	0.0000	−0.7050	3.4750	0.0000	0.0000	6,950†
x_3	8.34	0.8108	0.0000	1.0000	0.5405	0.8108	0.0000	−0.2703	270.27
x_6	0.00	2.4189	0.0000	0.0000	4.4459*	3.9189	1.0000	−2.9730	972.97
x_2	7.30	−0.4459	1.0000	0.0000	−0.2973	−0.9459	0.0000	0.6486	1,351.35
$z_j - c_j$		−1.7332	0.0000	0.0000	−1.8422	−0.1432	0.0000	2.4811	12,119†
x_3	8.34	0.5167*	0.0000	1.0000	0.0000	0.3343	−0.1216	0.0912	151.98
x_4	4.18	0.5441	0.0000	0.0000	1.0000	0.8815	0.2249	−0.6687	218.84
x_2	7.30	−0.2842	1.0000	0.0000	0.0000	−0.6839	0.0669	0.4498	1,416.41
$z_j - c_j$		−0.7310	0.0000	0.0000	0.0000	1.4805	0.4143	1.2492	12,522†
x_1	5.24	1.0000	0.0000	1.9353	0.0000	0.6471	−0.2353	0.1765 ‡	294.11
x_4	4.18	0.0000	0.0000	−1.0529	1.0000	0.5294	0.3529	−0.7647	58.82
x_2	7.30	0.0000	1.0000	0.5500	0.0000	−0.5000	0.0000	0.5000	1,500.00
$z_j - c_j$		0.0000	0.0000	1.4146	0.0000	1.9537	0.2424	1.3782 ¶	12,737†

* Pivot element.

† Value of objective function for that iteration.

‡ Inverse for the basis set of vectors.

¶ Dual solution, which is defined later.

corresponding to a variable that is not in the basis, the interpretation is that this variable could be entered into the program without changing the profit, and hence we have an alternate solution. A summary of the five tableaux is given on the preceding page.

3-9 COMMENTS

It was pointed out earlier that a brute-force method of finding this optimal solution would require the solution of 35 sets of three simultaneous linear equations. The method discussed above required the solution of only five such sets, including the one with which we started. The initial set of basic variables was (x_5, x_6, x_7), and in turn we evaluated the following sets: (x_3, x_6, x_7); (x_3, x_6, x_2); (x_3, x_4, x_2); and (x_1, x_4, x_2). The sequence of entering

Macro flow chart for simplex method

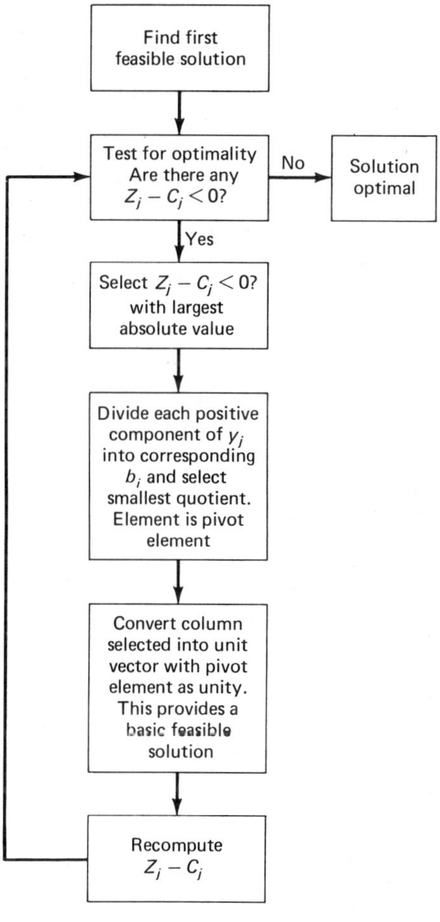

the variables into the solution is quite efficient, since only one variable was programmed for use and later removed from the program. The optimal solution resulted in noninteger answers, which is not always satisfactory. These can be eliminated by methods of integer programming, but we will not discuss these in this text. The sequence of operations used to obtain the optimal solution is shown in the form of a flow chart on p. 71.

In the summary of the tableaux given on p. 70, several things are highlighted and marked with asterisks: the value of the objective function at each iteration; the pivot elements; the inverse of the basis set of vectors for the optimal tableau; and the dual solution, which is explained later.

3-10 COMPUTER SOLUTION

The detailed solution just given requires a large number of computations, even though it is a relatively small problem. No one in his right mind would ordinarily go to the trouble of making all of these computations on a manual basis or even with the assistance of a desk calculator if an electronic computer is available. The real purpose of making the computations in such detail is to gain sufficient understanding to make proper interpretations.

If one has access to an electronic computer, he most likely will also have access to one or more programs which will solve a rather standard linear programming problem. In using such programs, it is of course necessary to know linear programming and enough about the computer system being used to be able to get such problems onto the computer, to be able to instruct the computer to perform the computations you want, and to get an output from the computer that will give you the essential information for the interpretation of the results. Only a modest knowledge of the computer is necessary for you to do this. Typically, the output consists of a full initial tableau, an intermediate summary of some kind showing the essential results of each iteration, and then a final tableau. Additional summaries are usually available to enable the user to make appropriate sensitivity analyses and to make a full interpretation of the solution.

3-11 COMMENTS ON LINEAR INDEPENDENCE

Referring to the final summary on p. 70, we can make several important observations. The basis variables in the optimal solution are x_1, x_4, and x_2. These were initially associated with the activity vectors

$$\begin{bmatrix} 1.5 \\ 1 \\ 1.5 \end{bmatrix}, \begin{bmatrix} 1 \\ 3.5 \\ 1 \end{bmatrix}, \text{ and } \begin{bmatrix} 1 \\ 5 \\ 3 \end{bmatrix},$$

respectively, but these have in the process of four iterations been transformed into the unit vectors

$$\begin{bmatrix}1\\0\\0\end{bmatrix}, \begin{bmatrix}0\\1\\0\end{bmatrix}, \text{ and } \begin{bmatrix}0\\0\\1\end{bmatrix}.$$

With the vectors in the form of unit vectors and the information from the final tableau, it is easy to write any one of the nonbasis vectors as a linear combination of these unit vectors, which we now designate as $\mathbf{y}_1$, $\mathbf{y}_4$, and $\mathbf{y}_2$. For example,

$$\begin{aligned}\mathbf{y}_5 &= 0.6471\mathbf{y}_1 + 0.5294\mathbf{y}_4 - 0.5000\mathbf{y}_2\\ &= 0.6471\begin{bmatrix}1\\0\\0\end{bmatrix} + 0.5294\begin{bmatrix}0\\1\\0\end{bmatrix} - 0.5000\begin{bmatrix}0\\0\\1\end{bmatrix}\\ &= \begin{bmatrix}0.6471\\0.5294\\-0.5000\end{bmatrix}\end{aligned}$$

which is read directly from the x_5 column in the final tableau. The linear combination relationship above holds in the initial tableau as well, but it is not as easy to see it there directly as it is from the initial tableau

$$\mathbf{a}_5 = \begin{bmatrix}1\\0\\0\end{bmatrix} = 0.6471\begin{bmatrix}1.5\\1\\1.5\end{bmatrix} + 0.5294\begin{bmatrix}1\\3.5\\1\end{bmatrix} - 0.5000\begin{bmatrix}1\\5\\3\end{bmatrix}$$

This representation, in vector form, is the equivalent of writing

$$x_5 = 0.6471x_1 + 0.5294x_4 - 0.5000x_2$$

This relationship will remain the same for each of the tableaux, but the actual combination is most easily seen when the transformation has been made so that $\mathbf{y}_1$, $\mathbf{y}_4$, and $\mathbf{y}_2$ are given as unit vectors. Not only is it easy to see that the vectors are linearly independent, but it is possible to write the linear combination by inspection. The property of linear independence is the key factor that allows the set of three unit vectors to span the space. As a matter of fact, any set of three vectors in the original matrix $\mathbf{A}$ are linearly independent and can be used to span the space. The particular choice in this case is dependent upon a number of factors, paramount of which is the fact that it also provides the optimal solution for the problem.

3-12 APPLICATIONS

There are many applications in which linear programming techniques are appropriate. *We are not yet prepared to handle the full range of linear programming problems, so the applications presented will be limited in scope to those problems in which the constraints are of the equal-to-or-less-than variety and the objective function is to be maximized.* As long as we restrict ourselves to that class of problems, we are in a position to obtain a complete algebraic solution by the simplex method as just described. Extensions to minimization problems and to a variety of $\leqq$, $\geqq$, and $=$ constraints will be made in the chapters that follow.

Manufacturing-Mix Problem

A small electrical manufacturing company manufactures two types of electrical switches in one of its departments, a two-pole switch and a three-pole switch. The net profit on a two-pole switch is 45 cents/switch and the net profit on a three-pole switch is 30 cents/switch. Each of the two-pole switches requires 2 min of machine time, whereas each of the three-pole switches requires only 1 min of machine time. The department has five machines, each of which provides 200 usable minutes per day.

Each of the switches requires 1 min of direct labor and two laborers are employed for this work, each providing 400 usable minutes of labor per day. The forecast recently made indicates that the company can sell up to 400 two-pole and 700 three-pole switches per day.

Formulate this problem to determine how many of each type of switch should be made to maximize net profit. Let

x_1 = number of two-pole switches to be made
x_2 = number of three-pole switches to be made

The function to be maximized is then

$$0.45x_1 + 0.30x_2$$

The constraints on the manufacturing are as follows:

Machine time	$2x_1 + x_2 \leqq 1000$
Labor time	$x_1 + x_2 \leqq 800$
Demand (two-pole)	$x_1 \leqq 400$
Demand (three-pole)	$x_2 \leqq 700$

By adding slacks x_3, x_4, x_5, and x_6, in order to the preceding constraints, our problem becomes:

Find

$$x_1, x_2, x_3, x_4, x_5, x_6 \geqq 0$$

such that

$$\begin{aligned} 2x_1 + x_2 + x_3 \qquad\qquad\qquad &= 1000 \\ x_1 + x_2 \qquad + x_4 \qquad\qquad &= 800 \\ x_1 \qquad\qquad\qquad + x_5 \qquad &= 400 \\ x_2 \qquad\qquad\qquad + x_6 &= 700 \end{aligned}$$

such that

$$z = 0.45x_1 + 0.30x_2 + 0x_3 + 0x_4 + 0x_5 + 0x_6 \text{ is a } maximum$$

The initial tableau used for solving this problem will then be:

Basis \ c_j		x_1 0.45	x_2 0.30	x_3 0	x_4 0	x_5 0	x_6 0	b 0000
x_3	00	2	1	1	0	0	0	1000
x_4	00	1	1	0	1	0	0	800
x_5	00	(1)	0	0	0	1	0	400
x_6	00	0	1	0	0	0	1	700
$z_j - c_j$		−0.45	−0.30	0	0	0	0	000

The successive tableaux are shown on page 76; the pivot element is marked in each case.

The optimal solution is thus to produce

$$x_1 = 200 \text{ two-pole switches}$$

$$x_2 = 600 \text{ three-pole switches}$$

with a profit of $270. The constraints on machine time and labor are both binding, but neither of the demand constraints is binding, since they are not able to make as many as the forecast says they would be able to sell.

Basis	c_j	x_1 0.45	x_2 0.30	x_3 0	x_4 0	x_5 0	x_6 0	b 0000
x_3	0	0	①	1	0	−2	0	200
x_4	0	0	1	0	1	−1	0	400
x_1	0.45	1	0	0	0	1	0	400
x_6	0	0	1	0	0	0	1	700
$z_j - c_j$		0	−0.30	0	0	0.45	0	180
x_2	0.30	0	1	1	0	−2	0	200
x_4	0	0	0	−1	1	①	0	200
x_1	0.45	1	0	0	0	1	0	400
x_6	0	0	0	−1	0	2	1	500
$z_j - c_j$		0	0	0.30	0	−0.15	0	240
x_2	0.30	0	1	−1	2	0	0	600
x_5	0	0	0	−1	1	1	0	200
x_1	0.45	1	0	1	−1	0	0	200
x_6	0	0	0	1	−2	0	1	100
$z_j - c_j$		0	0	0.15	0.15	0	0	270

3-13 USE OF THE OBJECTIVE FUNCTION AS AN ADDITIONAL CONSTRAINT

In the solution of the problem that was given in great detail at the beginning of this chapter, the $z_j - c_j$ row was calculated by $\mathbf{c_B y}_j - c_j$. An alternative way of performing the computations is to treat the $z_j - c_j$ row in the same way as all other rows are treated (i.e., treat it as a constraint).

The reasoning behind this can be seen as follows. First write the objective function as

$$z = 5.24x_1 + 7.30x_2 + 8.34x_3 + 4.18x_4$$

which is equivalent to writing

$$z - 5.24x_1 - 7.30x_2 - 8.34x_3 - 4.18x_4 - 0x_5 - 0x_6 - 0x_7 = 0$$

This constraint is now added to the original constraints so that the problem can now be stated:

Find

$$x_1, x_2, x_3, x_4, x_5, x_6, x_7 \geqq 0$$

such that

$$\begin{aligned}
1.5x_1 + x_2 + 2.4x_3 + x_4 + x_5 &= 2000 \\
x_1 + 5x_2 + x_3 + 3.5x_4 + x_6 &= 8000 \\
1.5x_1 + 3x_2 + 3.5x_3 + x_4 + x_7 &= 5000 \\
z - 5.24x_1 - 7.30x_2 - 8.34x_3 - 4.18x_4 - 0x_5 - 0x_6 - 0x_7 &= 0000
\end{aligned}$$

such that z is a maximum. For our initial program or solution,

$$\begin{aligned}
x_5 &= 2000 \\
x_6 &= 8000 \\
x_7 &= 5000 \\
z &= 0000
\end{aligned}$$

$$x_1 = x_2 = x_3 = x_4 = 0$$

In dealing with the problem in this form, we have an additional constraint as well as the additional variable z. By referring to the z constraint, one can see that increasing the value of x_3 while holding the remainder of the variables not presently in the solution at the zero level will require that z be increased, since the coefficient of x_3 in the z constraint is negative. Considering the basic feasible solutions of the system of four equations in eight unknowns shown above, one wishes to find that solution which has z as one of the basic variables, and at the same time makes z as large as possible.

Since z must always be a basic variable, it will always appear in the last equation only, and its value will be found in the b-vector column in the same way as it was for the other variables. The initial tableau can then be written as follows, where there is no need to have a row designated as c_j. Furthermore, it is not necessary to have a row designated as $z_j - c_j$ because the equivalent is in the z constraint row.

Initial tableau

		z	x_1	x_2	x_3	x_4	x_5	x_6	x_7	b
$\mathbf{x_B}$	$\mathbf{c'_B}$									
x_5	0.00	0	1.5000	1.0000	2.4000	1.0000	1.0000	0.0000	0.0000	2000
x_6	0.00	0	1.0000	5.0000	1.0000	3.5000	0.0000	1.0000	0.0000	8000
x_7	0.00	0	1.5000	3.0000	3.5000	1.0000	0.0000	0.0000	1.0000	5000
z	1.00	1	−5.2400	−7.3000	−8.3400	−4.1800	0.0000	0.0000	0.0000	0000

The indicators for the entering variable appear in the last row, as before (i.e., −8.34, being the largest negative, indicates that x_3 is the entering variable). Since z will always remain in the basis, the only candidates for the pivot point correspond to the first three constraints. The selection is the same as before, and 2.4 becomes the pivot element. The operations for the first three rows are as before, but now the fourth row is treated in the same manner as the other rows. In the initial tableau the vector corresponding to x_3 is

$$\begin{bmatrix} 2.4 \\ 1 \\ 3.5 \\ -8.34 \end{bmatrix}$$

and it will be transformed into the unit vector

$$\begin{bmatrix} 1 \\ 0 \\ 0 \\ 0 \end{bmatrix}$$

The resulting second tableau is:

Second tableau

		z	x_1	x_2	x_3	x_4	x_5	x_6	x_7	b
$\mathbf{x_B}$	$\mathbf{c'_B}$									
x_3	8.34	0	0.6250	0.4167	1.0000	0.4167	0.4167	0.0000	0.0000	833.33
x_6	0.00	0	0.3750	4.5833	0.0000	3.0833	−0.4167	1.0000	0.0000	7166.67
x_7	0.00	0	−0.6875	1.5417	0.0000	−0.4583	−1.4583	0.0000	1.0000	2083.33
z	1.00	1	−0.0275	−3.8250	0.0000	−0.7050	3.4750	0.0000	0.0000	6950

It should be noticed that this is essentially the same as the second tableau previously used, with the following slight modifications:

1. There is a new column, labeled z, which remains the same unit vector throughout all iterations. Thus it is not really necessary to have it in the tableau.
2. The last row (corresponding to the basic variable z) is exactly the same as the $z_j - c_j$ row, and since it incorporates the value of c_j, the c_j row is not necessary.

The components in the last row, since they are really the $(z_j - c_j)$'s, provide a test for improvement.

3-14 ALTERNATIVE LOGIC

The logic of the test for improvement can be seen in another way by writing the equation represented by the last row of the tableau. In the second tableau, for example, this row can be written

$$z - 0.0275x_1 - 3.8250x_2 - 0.7050x_4 + 3.4750x_5 = 6950$$

and solving for z,

$$z = 0.0275x_1 + 3.8250x_2 + 0.7050x_4 - 3.4750x_5 + 6950$$

At this point in the solution, x_1, x_2, x_4, and x_5 are nonbasic variables and carry the value of zero. Consequently, the value of z is 6950. It can be seen that increasing x_1, x_2, or x_4 will increase the value of the objective function, because the coefficients of these variables are all positive. The selection of x_2 as the next variable to enter the basis is made because its coefficient has the largest positive value and hence indicates the largest per unit improvement. It can also be seen that entering the variable x_5 will cause a decrease in the value of z since its coefficient is negative.

Since the two methods of handling the objective function are seen to be equivalent, let us return to the original way of writing the tableau, since it has one less column. In future problems, we shall place the c_j row at the top of the first tableau for reference purposes and not repeat it. The last row in each tableau will be labeled $z_j - c_j$, and it will be handled in the same manner as any other row as far as computation is concerned.

Earlier in this chapter it was pointed out that either $c_j - z_j$ or $z_j - c_j$ could be used as the test for improvement. The development described above gives us reason for using it in the form $z_j - c_j$.

As pointed out previously, it is quite easy to write any one of the nonbasic variables as a linear combination of the basic variables. This was illustrated for the variable x_5 on p. 73. This process may be applied to any of the tableau.

The $z_j - c_j$ row has as its primary purpose that of testing for possible improvement in the solution, but it also pays an important part in sensitivity analysis, as will be described in a later chapter.

3-15 BASIS INVERSE

In Chap. 2 a method for inverting a matrix was developed. The method used was that of annexing an identity matrix to the original matrix. Transformations were then made to reduce the original matrix to an identity matrix. The matrix which then appears in place of the original identity matrix is the inverse of the subject matrix.

Applying this principle to the optimal tableau on p. 69, we find that the matrix in columns x_5, x_6, and x_7 is the inverse of some matrix. To find the matrix whose inverse it represents, we must locate the columns of the identity matrix in this final tableau and then the values which were originally in these positions. By reference to the definition of unit vectors on p. 22, we see that $\mathbf{e}_1$ is under x_1, $\mathbf{e}_2$ is under x_4, and $\mathbf{e}_3$ is under x_2. The original vectors in these positions in that same order are given by

$$\mathbf{B} = \begin{bmatrix} 1.5 & 1 & 1 \\ 1 & 3.5 & 5 \\ 1.5 & 1 & 3 \end{bmatrix}$$

and therefore we find in columns x_5, x_6, and x_7 that

$$\mathbf{B}^{-1} = \begin{bmatrix} 0.6471 & -0.2353 & 0.1765 \\ 0.5294 & 0.3529 & -0.7647 \\ -0.5000 & 0.0000 & 0.5000 \end{bmatrix}$$

Consequently, the solution for the basic variables is

$$\mathbf{x_B} = \begin{bmatrix} x_1 \\ x_4 \\ x_2 \end{bmatrix} = \mathbf{B}^{-1}\mathbf{b} = \begin{bmatrix} 0.6471 & -0.2353 & 0.1765 \\ 0.5294 & 0.3529 & -0.7647 \\ -0.5000 & 0.0000 & 0.5000 \end{bmatrix} \begin{bmatrix} 2000 \\ 8000 \\ 5000 \end{bmatrix} = \begin{bmatrix} 294.11 \\ 58.82 \\ 1500.00 \end{bmatrix}$$

Since each of the columns of the original $\mathbf{A}$ matrix can be handled in the same way as the $\mathbf{b}$ vector above, the complete final tableau can be obtained by evaluating $\mathbf{B}^{-1}\mathbf{A}$:

$$\begin{bmatrix} 0.6471 & -0.2353 & 0.1765 \\ 0.5294 & 0.3529 & -0.7647 \\ -0.5000 & 0.0000 & 0.5000 \end{bmatrix} \begin{bmatrix} 1.5 & 1 & 2.4 & 1 & 1 & 0 & 0 & 2000 \\ 1 & 5 & 1 & 3.5 & 0 & 1 & 0 & 8000 \\ 1.5 & 3 & 3.5 & 1 & 0 & 0 & 1 & 5000 \end{bmatrix}$$

which gives the final tableau exclusive of the $z_j - c_j$ row. These relationships will be used frequently in later chapters when we discuss sensitivity analysis.

3-16 SUMMARY

The entire chapter has been devoted to the solution of a straightforward maximization problem with equal-to-or-less-than constraints. Details have been provided so that the reader can confirm each of the computations. The flow chart on p. 71 shows the actual sequence of steps and decisions that are made. The summary of all the tableaux on p. 70 provides a running record of the progress as the solution moves toward the optimum. With the information provided, one should now be able to solve any linear programming model that requires maximization and has nothing but equal-to-or-less-than constraints.

There are a number of facts that should be pointed out that may have been missed by the reader. Most of these can be seen by reference to the summary of all the tableaux.

1. Each of the tableaux contains an identity matrix.
2. The three-component **b** vector has each of its components equal to or greater than zero in each of the tableaux, making all solutions feasible.
3. Even though the selection of the incoming vector was made by selection on the basis of the negative $z_j - c_j$ with the largest absolute value, improvement could have been made by selecting any vector with a negative $z_j - c_j$.
4. After the selection of the vector to be entered, the pivot point was determined by considering only positive components of the entering vector.
5. The solution progressed from an initial basic feasible solution to another basic feasible solution and to another basic feasible solution until an optimal basic feasible solution was reached.
6. According to the way in which the algorithm was set up, each decision was clear. There were no ties. Had there been equal values for two different $z_j - c_j$'s, either could have been used. Had there been a tie in the choice of a pivot point, either could have been used. The latter would lead to a degenerate solution at that next stage, which might cause modest but not insurmountable difficulties. Degeneracy is a fact of life in real-world linear programming problems and of course must be handled.
7. When changing basic variables in the process of the solution, only one variable was entered at a time. An algorithm for entering two at a time could be developed, but there seems to be no advantage in doing so, since the method is more than twice as complicated as entering one at a time, and besides the selection of the appropriate two vectors to be entered is not clear.
8. One cannot emphasize too strongly the fact that the final tableau can be obtained by multiplying $\mathbf{B}^{-1}$, as found from the optimal tableau, times the original initial tableau.

EXERCISES

The purposes of the exercises in this chapter is to provide a means of mastering the basic elements of the simplex method for the case in which the objective function is to be *maximized* and all constraints are of the equal-to-or-less-than variety.

1 Solve the following problem by the simplex method.

Find

$$x_1, x_2 \geqq 0$$

such that

$$2x_1 - x_2 \leqq 8$$

$$2x_1 + x_2 \leqq 12$$

$$x_1 + 2x_2 \leqq 16$$

and such that

$$z = 4x_1 + 3x_2 \text{ is a } \textit{maximum}$$

Show all the tableaux. Carry out the computations using fractions rather than rounded decimals so that the answers will be exact (i.e., no rounding error).

2 Solve the following problem by the simplex method.

Find

$$x_1, x_2, x_3 \geqq 0$$

such that

$$3x_1 + 2x_2 - x_3 \leqq 610$$

$$-x_1 + 6x_2 + 3x_3 \leqq 125$$

$$-2x_1 + x_2 + 0.5x_3 \leqq 420$$

and such that

$$z = 2x_1 + 5x_2 + 8x_3 \text{ is a } \textit{maximum}$$

Show all the tableaux. Carry out the computations using fractions rather than rounded decimals so that the answers will be exact.

3 Solve by the simplex method the product-mix problem stated in Exercise 1, Chap. 2 (p. 55).

4 Solve by the simplex method the gasoline blending problem stated as Exercise 2, Chap. 2 (p. 55).

In writing the mathematical statement for this problem, the normal method will probably lead to two $\geqq$ constraints. However, to solve this problem with our present knowledge, it is necessary to have all $\leqq$ constraints. The abovementioned constraints can easily be converted to $\leqq$ constraints, since the right-hand sides for both cases are zeros.

In carrying out the solution to this problem, be especially careful. Use only positive elements in a vector as candidates for pivot points, once the entering vector has been determined. Furthermore, be sure that you understand that when you divide a number into zero, the result is zero, which is less than any positive quotient.

5 In Exercise 1 find $\mathbf{B}^{-1}$, the inverse of the basis set of vectors for the optimal solution. From this inverse, obtained from the final tableau, find the optimal solution by computing

$$\mathbf{x_B} = \mathbf{B}^{-1}\mathbf{b}$$

where $\mathbf{b}$ is the vector of right-hand-side elements in the initial tableau.

6 In Exercise 2 find $\mathbf{B}^{-1}$, the inverse of the basis set of vectors from the final tableau. From this inverse, obtained from the final tableau, find the optimal solution by computing

$$\mathbf{x_B} = \mathbf{B}^{-1}\mathbf{b}$$

where $\mathbf{b}$ is the vector of right-hand-side elements in the initial tableau.

7 For Exercise 4 find $\mathbf{B}^{-1}$ from the optimal tableau, and show that by multiplying this $\mathbf{B}^{-1}$ by the entire initial tableau, you can obtain the entire final tableau.

8 Solve Example 1-2 (p. 5), by the simplex method.

9 Solve Exercise 1, Chap. 1 (p. 11), by the simplex method.

10 Solve Exercise 3, Chap. 1 (p. 12), by the simplex method.

CHAPTER

FOUR

GENERALIZED LINEAR PROGRAMMING PROBLEMS

4-1 EQUAL-TO-OR-GREATER-THAN AND EQUAL-TO CONSTRAINTS

The illustrations and problems given in Chap. 3 were rather simple and straightforward. In every case the constraints were of the equal-to-or-less-than type. The problems were initially written so that the components of the right-hand side were always equal to or greater than zero. One of the problems actually had zeros for some of the right-hand sides, but other than that there were no complications. Furthermore, at each decision point in the process of solution, there was no doubt about the decision or the ability to carry it out. Such is not always the case, so many of the more common exceptions will be considered.

Consider the more general problem of finding $\mathbf{x} \geqq \mathbf{0}$, subject to a set of linear restrictions, some of which are $\leqq$, some $\geqq$, and some $=$'s, and requiring the maximization of $z = \mathbf{cx}$. To handle this more generalized problem, we shall make the following adaptations:

1. Assume that the constraints have been written in the form $\mathbf{Ax}(\leqq, \geqq, =)\mathbf{b}$, with each of the constraints written so that each component of $\mathbf{b}$ is nonnegative. It may be necessary to change the sense of some of the inequalities to accomplish this.
2. For each $\leqq$ inequality, a new variable, called *slack*, is added; the inequality then becomes an equation. The slack variable that appears in the

equation will appear in no other equation, and therefore the vector corresponding to this variable can be represented by a column of the identity matrix. The slack variable fills in the difference between the left-hand side of the inequality and the **b** value and will always be nonnegative.

3. For each $\geqq$ inequality, a new variable, called *surplus* or negative slack, is subtracted—the inequality becomes an equation. The surplus represents the amount by which the left-hand side of the inequality exceeds the **b** value, and it will always be nonnegative, since $\mathbf{b} \geqq \mathbf{0}$.

Both the slack in 2, and the surplus mentioned in 3, will be assigned a zero coefficient in the objective function, since in general neither of them will contribute anything to the function being optimized. Having added slack and subtracted surplus variables where needed, we have converted all the constraints to equations with the right-hand side equal to a nonnegative number. In the case of the $\leqq$ inequality, the slack variable also supplies a column for the identity matrix in the initial tableau. This is not true in the case of the $\geqq$ inequality, nor is it true for the constraints which were initially written in equation form.

Since the simplex method of solution relies upon having an identity matrix present at all times, it is necessary to supply the appropriate column if such is not already present.

4. For each of the constraints (now in equation form) corresponding to original $\geqq$ inequalities as well as for the original equations themselves, a new variable must be added. This new variable will have a coefficient of $+1$. Since the new variable has been added for the express purpose of obtaining a starting point for the simplex process, it is designated as an "artificial" variable, and because of this it can never be allowed to appear in the optimal solution. In order to make certain of this, an extremely high penalty is attached to the variable. This is done by assigning a very large negative number as its coefficient in the objective function. In much of the literature, you will find this coefficient designated by $-M$ in a maximization problem, and we shall do the same at the outset.

4-2 USE OF ARTIFICIAL VARIABLES

In order to understand the idea involved with "equal-to" and "equal-to-or-greater-than" constraints and the associated variables, let us carry an illustration far enough to fully explain the ideas involved. Consider the following example.

A small electronics supplier has as a part of his business the assembly of do-it-yourself kits for making hand-held calculators, tape recorders, and a small stereo set. He assembles the kits over the weekend for sale during the next week. There are some restrictions in carrying out the business. These are:

The time available for the assembly of the kits limits the total number to be assembled to 500 units regardless of which kit is being assembled.

Marketing reports indicate that in order to keep potential customers reasonably happy, he must have at least 100 kits in calculators and recorders together.

The calculator seems to be the most popular kit, and therefore the number of these kits must exceed the number of recorders and stereos together. The amount of the excess has been arbitrarily set to 120 units.

The profit from the assembly of the calculator kit is \$5; from the tape recorder, \$10; and from the stereo, \$15.

On the assumption that sales itself is no problem as long as the restrictions above are satisfied, how many of each of the kits should be assembled over a weekend to maximize the profit from the operation?

In forming the mathematical model, let

x_1 = number of calculators to be assembled
x_2 = number of recorders to be assembled
x_3 = number of stereos to be assembled

The problem statement is, then:

Find

$$x_1, x_2, x_3 \geqq 0$$

such that

$$\begin{aligned} x_1 + x_2 + x_3 &\leqq 500 \\ x_1 + x_2 \quad\quad &\geqq 100 \\ x_1 - x_2 - x_3 &= 120 \end{aligned}$$

and such that

$$z = 5x_1 + 10x_2 + 15x_3 \text{ is a } \textit{maximum}$$

The example above has been selected because it embodies all of the possible characteristics of a maximization problem (i.e., it has one each of every possible kind of constraint). If we know how to solve this problem, it should be possible to solve every possible maximization problem.

As indicated in the first part of this chapter, a slack variable must be added to the $\leqq$ constraint; a surplus must be subtracted from and an artificial variable must be added to the $\geqq$ constraint; and an artificial variable must be added to the = constraint. If these variables are provided in subscript order as follows—the surplus for the second constraint, slack for the first constraint, the artificial for the second constraint, and the artificial for the third constraint—the constraints for the problem become

$$
\begin{aligned}
x_1 + x_2 + x_3 \qquad\quad + x_5 \qquad\qquad &= 500 \\
x_1 + x_2 \qquad - x_4 \qquad + x_6 \qquad &= 100 \\
x_1 - x_2 - x_3 \qquad\qquad\qquad + x_7 &= 120
\end{aligned}
$$

Just as we have found in all our linear programming problems, it is very simple to find a basic feasible solution, in this case with x_5, x_6, and x_7 as the basic variables. The variables x_6 and x_7 are artificial and as such cannot remain in the solution. Therefore, a heavy penalty will be attached to each of these variables, and it is done by giving each a coefficient of $-M$ in which M itself is a very large number. With these additions, we now have:

Initial tableau

c_j		5	10	15	0	0	$-M$	$-M$	000
Variables		x_1	x_2	x_3	x_4	x_5	x_6	x_7	b
$\mathbf{x_B}$	$\mathbf{c'_B}$								
x_5	00	1	1	1	0	1	0	0	500
x_6	$-M$	(1)	1	0	-1	0	1	0	100
x_7	$-M$	1	-1	-1	0	0	0	1	120
$z_j - c_j$		$-2M - 5$	-10	$M - 15$	M	0	0	0	$-220M$
		↑					↓		

It will be noted that by adding the necessary additional variables in the order in which they were added, we find that we conveniently have the necessary identity matrix in the right-hand portion of the initial tableau.

From this tableau it can be seen that the initial basic feasible solution is

$$x_5 = 500$$

$$x_6 = 100$$

$$x_7 = 120$$

$$x_1 = x_2 = x_3 = x_4 = 0$$

with the value of the objective function being $-220M$.

For the initial tableau, the $z_j - c_j$ is computed by using $\mathbf{c_B a}_j - c_j$ and, after that, it is treated in the same manner as any other row.

To determine the entering variable, we must find the negative in the $z_j - c_j$ row with the largest absolute value. The only negative terms are -10 and $(-2M - 5)$. Since M can be assigned an arbitrarily large value, $-2M - 5$ obviously has a larger absolute value than -10. The entering variable will be x_1 and the pivot element is chosen in the regular fashion so that the variable x_6 will be removed from the basis, and the result is:

Second tableau

c_j		5	10	15	0	0	$-M$	$-M$	000
Variables		x_1	x_2	x_3	x_4	x_5	x_6	x_7	b
$\mathbf{x_B}$	$\mathbf{c'_B}$								
x_5	00	0	0	1	1	1	-1	0	400
x_1	5	1	1	0	-1	0	1	0	100
x_7	$-M$	0	-2	-1	①	0	-1	1	20
$z_j - c_j$		0	$2M - 5$	$M - 15$	$-M - 5$	0	$2M + 5$	0	$-20M + 500$
					↑			↓	

Since $-M - 5$ in the $z_j - c_j$ is the only negative term when M is assigned an arbitrarily large value, variable x_4 will be entered next and x_7 will be removed from the basis.

The result is:

Third tableau

c_j		5	10	15	0	0	$-M$	$-M$	000
Variables		x_1	x_2	x_3	x_4	x_5	x_6	x_7	b
$\mathbf{x_B}$	$\mathbf{c'_B}$								
x_5	00	0	2	②	0	1	0	-1	380
x_1	5	1	-1	-1	0	0	0	1	120
x_4	00	0	2	1	1	0	-1	1	20
$z_j - c_j$		0	-15	-20	0	0	M	$M + 5$	600
				↑		↓			

The next variable to be entered is x_3, and then variable x_5 will be removed from the basis. The result is:

Optimal tableau

c_j		5	10	15	0	0	$-M$	$-M$	000
Variables		x_1	x_2	x_3	x_4	x_5	x_6	x_7	b
$\mathbf{x_B}$	$\mathbf{c'_B}$								
x_3	15	0	1	1	0	0.5	0	-0.5	190
x_1	5	1	0	0	0	0.5	0	0.5	310
x_4	00	0	-1	0	1	0.5	-1	0.5	210
$z_j - c_j$		0	5	0	0	10	M	$M - 5$	4400

It should be noticed that there are no negative $(z_j - c_j)$'s and hence the tableau is optimal. Furthermore, since artificial variable x_6 was removed from the basis in the first iteration, it could have been stricken from the tableau, since it will never be a candidate for reentry into the basis. The same can be said for variable x_7 after the second iteration. These variables have been carried throughout the process, for reasons that will become clear later.

The optimal solution to this problem is then

$$x_3 = 190$$

$$x_1 = 310$$

$$x_4 = 210$$

$$x_2 = x_5 = x_6 = x_7 = 0$$

and the value of the objective function is 4400. The two artificial variables really need not be given as a part of the solution, since they were not even in the original statement of the problem.

4-3 MEANING OF THE ARTIFICIAL VARIABLES

When one has a new concept to explain, one often searches for an analogy to assist in the explanation. I like to compare it with a situation with which I have become familiar in recent years. I suggest that an "artificial variable" is very much like a "handle" on a bowling ball. Many people laugh when I say this, since most people have probably never seen a bowling ball with a handle. Such a thing does exist, however, and has a very real and special purpose. It has been made for those people who cannot put their fingers into a regular bowling ball for some reason or other, usually because of arthritis, and need some method of getting the ball out on the alley. Once that has been accomplished, the handle is not needed and therefore might just as well disappear.

The diagram shows the construction of such a bowling ball. The handle is spring-loaded. When you first see the ball, the handle forms a part of the contour of the ball. There is a small lip which enables one to pull the handle out to the position shown below. When the ball is thrown, the handle snaps back into the ball before the ball hits the alley and therefore it rolls exactly like any other bowling ball. The handle has served its purpose and is therefore no longer needed—it acts just like an artificial variable.

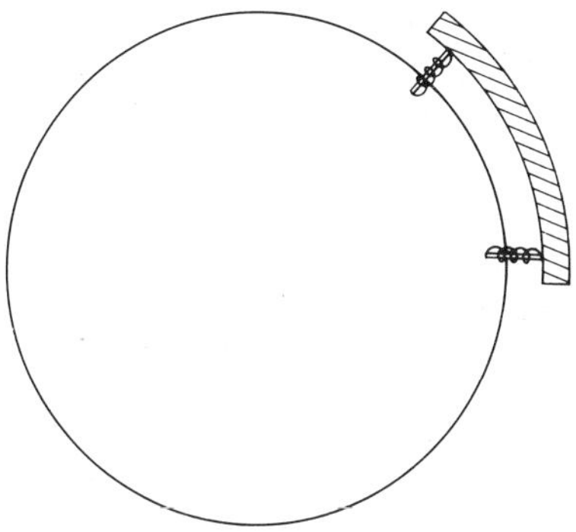

Of course, the initial tableau contains an identity matrix, since variables were added to provide just that. This identity matrix is actually in columns x_5, x_6, and x_7. From our developments in Chap. 3, it is known that the matrix in those columns in the optimal tableau is the inverse of the basis set of vectors. The basis set of vectors in the final tableau is in columns x_3, x_1, and x_4. Therefore,

$$\mathbf{B} = \begin{array}{c} \begin{matrix} x_3 & x_1 & x_4 \end{matrix} \\ \begin{bmatrix} 1 & 1 & 0 \\ 0 & 1 & -1 \\ -1 & 1 & 0 \end{bmatrix} \end{array} \qquad \text{and} \qquad \mathbf{B}^{-1} = \begin{bmatrix} 0.5 & 0 & -0.5 \\ 0.5 & 0 & 0.5 \\ 0.5 & -1 & 0.5 \end{bmatrix}$$

This inverse may be used to transform the original tableau into the final tableau.

4-4 TWO-PHASE METHOD OF HANDLING ARTIFICIAL VARIABLES

The problem whose solution was just obtained was handled in the regular fashion after the artificial variables had been added. There is a slight complication in this method, in that one must carry the value of M as a very large number without specifying exactly what the number is. If a specific number were assigned to it, the number might not take care of all situations. Its purpose is to provide a penalty to drive out the artificial variables, since they really cannot be a part of the solution to a real-world problem. A device to get around this difficulty is incorporated into what is known as the *two-phase method*. The first phase consists of driving out the artificial variables in order to obtain a basic feasible solution for the real variables of the problem. The second phase consists of using this basic feasible solution as a starting point from which to develop the optimal solution by the regular simplex method.

Let us now consider the two-phase method for the problem described above. The initial tableau will be exactly as before, with the exception that the coefficients of the objective function will all be zeros except for the artificial variables, in which case they will all be -1. Using this method, the first three tableaux will be as follows:

Initial tableau (phase I)

c_j		0	0	0	0	0	-1	-1	0000
Variables		x_1	x_2	x_3	x_4	x_5	x_6	x_7	b
$\mathbf{x_B}$	c'_B								
x_5	00	1	1	1	0	1	0	0	500
x_6	-1	(1)	1	0	-1	0	1	0	100
x_7	-1	1	-1	-1	0	0	0	1	120
$z_j - c_j$		-2	0	1	1	0	0	0	-220
		↑					↓		

Second tableau (phase I)

c_j		0	0	0	0	0	-1	-1	0000
Variables		x_1	x_2	x_3	x_4	x_5	x_6	x_7	b
$\mathbf{x_B}$	c'_B								
x_5	00	0	0	1	1	1	-1	0	400
x_1	00	1	1	0	-1	0	1	0	100
x_7	-1	0	-2	-1	(1)	0	-1	1	20
$z_j - c_j$		0	2	1	-1	0	2	0	-20
					↑			↓	

Third tableau (phase I)

c_j		0	0	0	0	0	−1	−1	0000
Variables		x_1	x_2	x_3	x_4	x_5	x_6	x_7	b
$\mathbf{x_B}$	$\mathbf{c'_B}$								
x_5	00	0	2	2	0	1	0	−1	380
x_1	00	1	−1	−1	0	0	0	1	120
x_4	00	0	−2	−1	1	0	−1	1	20
$z_j - c_j$		0	0	0	0	0	1	1	000

This tableau is optimal for the first phase of the problem. This fact is indicated in two different ways: first, the $z_j - c_j$ row has no negatives, and second, the objective function has been driven to zero, indicating that no artificial variables are present, which was the purpose of phase I. We thus have the following basic feasible solution in terms of the original variables:

$$x_5 = 380$$
$$x_1 = 120$$
$$x_4 = 20$$
$$x_2 = x_3 = x_6 = x_7 = 0$$

This solution must now be tested for optimality in terms of the original objective function. For this purpose we create an initial tableau, without the variables x_6 and x_7, with the original coefficients of the objective function.

Initial tableau (phase II)

c_j		5	10	15	0	0	000
Variables		x_1	x_2	x_3	x_4	x_5	b
$\mathbf{x_B}$	$\mathbf{c'_B}$						
x_5	00	0	2	(2)	0	1	380
x_1	5	1	−1	−1	0	0	120
x_4	00	0	−2	−1	1	0	20
$z_j - c_j$		0	−15	−20	0	0	600
				↑		↓	

Second tableau (phase II)

c_j		5	10	15	0	0	000
Variables		x_1	x_2	x_3	x_4	x_5	b
$\mathbf{x_B}$	$\mathbf{c'_B}$						
x_3	15	0	1	1	0	0.5	190
x_1	5	1	0	0	0	0.5	310
x_4	00	0	−1	0	1	0.5	210
$z_j - c_j$		0	5	0	0	10	4400

This tableau is optimal since the $z_j - c_j$ row contains no negatives. The solution is the same as before, of course. The two-phase method has the advantage that it can be easily programmed for the computer. Changing the coefficients of the objective function in the middle of the solution offers no difficulties.

4-5 MINIMIZATION PROBLEMS

Up to this point we have provided the means to solve any maximization problem for all possible types of constraints. The problem whose solution was just completed had one $\leqq$ constraint, one $\geqq$ constraint, and one $=$ constraint. The only additional thing that can exist is to have a larger problem and have more of each of the various types of constraints, but this adds nothing new—it is just more of the same.

The only thing new that really needs to be handled is to have a problem in which the objective function is to be minimized. To solve this type of problem, we shall convert a minimization problem to a maximization problem and, after solving that problem, convert our solution back to the solution of the original problem. How can this be done? Very simply. If $Z = \mathbf{cx}$ is to be minimized, the same solution will be obtained if $z = -\mathbf{cx}$ is maximized. That is, all that needs to be done is to change the signs of the coefficients of the objective function and solve that problem by the methods we have already developed. The solution vector $\mathbf{x}$ will be exactly the same, so in order to present the solution to the original problem, one needs only to reverse the sign of the objective function. What could be simpler?

We are now in position to solve a linear programming problem of any kind as long as we have sufficient time to carry out the computations. There are occasionally some complications that appear in the process of carrying out the solution. A number of these will be explained.

4-6 SOLUTION EXAMPLE INVOLVING MINIMIZATION AND A MIXTURE OF CONSTRAINTS

Let us now consider the problem given as Example 1-3 (p. 6). The solution for this problem was obtained by the graphical method at that point. The mathematical statement of the problem was:

Find

$$x_1, x_2 \geqq 0$$

such that

$$\begin{aligned} x_1 \qquad\quad &\leqq 3 \\ x_2 &\leqq 4 \\ x_1 + \ x_2 &\leqq 5 \\ x_1 + 5x_2 &\geqq 5 \\ 5x_1 + 2x_2 &\geqq 10 \end{aligned}$$

and such that

$$Z = 3x_1 + 2x_2 \text{ is a } \textit{minimum}$$

In order to solve this problem, we shall convert it to an equivalent problem

Find

$$x_1, x_2 \geqq 0$$

such that

$$\begin{aligned} x_1 \qquad\quad &\leqq 3 \\ x_2 &\leqq 4 \\ x_1 + \ x_2 &\leqq 5 \\ x_1 + 5x_2 &\geqq 5 \\ 5x_1 + 2x_2 &\geqq 10 \end{aligned}$$

and such that

$$z = -3x_1 - 2x_2 \text{ is a } \textit{maximum}$$

The solution to this problem requires that we add slack variables to each of the first three constraints and a surplus and an artificial variable to each of the last two constraints. From a theoretical point of view, it makes no difference which of these is added first, but note that the order in which I have added them provides for the identity matrix in the last five columns.

Summary

c_j **(Ph. I)** c_j		−3 0	−2 0	0 0	0 0	0 0	0 0	0 0	** −1	** −1	000 000
Variables		x_1	x_2	x_3	x_4	x_5	x_6	x_7	x_8	x_9	b
x_5	00	1	0	0	0	1	0	0	0	0	3
x_6	00	0	1	0	0	0	1	0	0	0	4
x_7	00	1	1	0	0	0	0	1	0	0	5
x_8	−1	1	(5)	−1	0	0	0	0	1	0	5
x_9	−1	5	2	0	−1	0	0	0	0	1	10
$z_j - c_j$		−6	−7	1	1	0	0	0	0	0	−15
x_5	00	1	0	0	0	1	0	0	0	0	3
x_6	00	$-\frac{1}{5}$	0	$\frac{1}{5}$	0	0	1	0	$-\frac{1}{5}$	0	3
x_7	00	$\frac{4}{5}$	0	$\frac{1}{5}$	0	0	0	1	$-\frac{1}{5}$	0	4
x_2	00	$\frac{1}{5}$	1	$-\frac{1}{5}$	0	0	0	0	$\frac{1}{5}$	0	1
x_9	−1	($\frac{23}{5}$)	0	$\frac{2}{5}$	−1	0	0	0	$-\frac{2}{5}$	1	8
$z_j - c_j$ **(Ph. II)**		$-\frac{23}{5}$	0	$-\frac{2}{5}$	1	0	0	0	$\frac{7}{5}$	0	−8
x_5	00	0	0	$-\frac{2}{23}$	$\frac{5}{23}$	1	0	0	$\frac{2}{23}$	$-\frac{5}{23}$	$\frac{29}{23}$
x_6	00	0	0	$\frac{5}{23}$	$-\frac{1}{23}$	0	1	0	$-\frac{5}{23}$	$\frac{1}{23}$	$\frac{77}{23}$
x_7	00	0	0	$\frac{3}{23}$	$\frac{4}{23}$	0	0	1	$-\frac{3}{23}$	$-\frac{4}{23}$	$\frac{60}{23}$
x_2	−2	0	1	$-\frac{5}{23}$	$\frac{1}{23}$	0	0	0	$\frac{5}{23}$	$-\frac{1}{23}$	$\frac{15}{23}$
x_1	−3	1	0	$\frac{2}{23}$	$-\frac{5}{23}$	0	0	0	$-\frac{2}{23}$	$\frac{5}{23}$	$\frac{40}{23}$
$z_j - c_j$		0	0	$\frac{4}{23}$	$\frac{13}{23}$	0	0	0	***	***	$-\frac{150}{23}$

At the top of the tableaux, both phase I and regular c_j's are given. The computations have been carried out with phase I c_j's, driving out the artificials x_8 and x_9 in two iterations. (In the third tableau, the $Z_j - C_j$ row has been calculated using original C_j's. It tests optimal with $z = -\frac{150}{23}$.) Two iterations were required to drive out the artificial variables, and when the regular coefficients of the objective function were substituted, it was shown that no improvement could be made. Therefore, for this problem, it was not necessary to do anything at all in phase II.

Since the value of the objective function for this maximizing problem was $-\frac{150}{23}$, the value of the objective function for the original minimizing problem is $\frac{150}{23}$. Therefore, the optimal solution is

$$x_1 = \frac{40}{23} \qquad \text{(chairs to be produced)}$$

$$x_2 = \frac{15}{23} \qquad \text{(tables to be produced)}$$

$$z = \frac{150}{23} \qquad \text{(minimum variable cost)}$$

which agrees completely with the graphical solution given in Chap. 1.

4-7 COMMENTS ON GENERAL SOLUTIONS

The methods discussed in Chaps. 3 and 4 provide the means of solving the standard linear programming problems once the algebraic statement of the problem has been made. It is entirely possible, however, that one may run into what one might consider some unusual circumstances, even though these situations can be handled routinely. The following gives a short description of some of the more common situations:

1. In the process of solving a linear program, it can happen that the $z_j - c_j$ row contains a negative term, indicating that improvement can be made, but that in searching the components in the column, only negatives can be found (i.e., it seems impossible to find a proper pivot element). Although it is not obvious, the fact is that this condition implies that no optimal can be found for the objective function. Then we have what is called an *unbounded solution*: an infinite amount of that variable can be entered into the solution with an ever-increasing value for the objective function. Most computer programs indicate this condition whenever it occurs—the program is terminated at this point.
2. As has been indicated, artificial variables are used to assist in solving certain types of problems. The artificial variables serve purely as an aid in getting a start on what would otherwise be a troublesome problem. Because of this artificial nature, an unusually heavy penalty is placed upon the use of the variable in order to drive it out of the solution. There are times, however, when an artificial variable appears in the optimal solution. The interpretation of this is that the real-world problem does not have a solution that meets all the constraints. Computer programs also indicate this condition.
3. In the material that was discussed in Chaps. 3 and 4, no particular emphasis was placed on making sure that the constraints were either nonredundant or consistent. The reason for omitting this emphasis was that neither inconsistencies nor redundancies cause any problems. One should take reasonable precautions to avoid problems, but having taken these precautions, no great trouble is encountered if you have failed to remove a redundancy or inconsistency. The redundancy may or may not be discovered in the normal process of solution. It is possible to get an optimal solution without realizing that one of the constraints is a linear combination of some of the others. An inconsistency is discovered at a point at which an entire row of the tableau, with the exception of the element in the **b** vector, has been reduced to zeros—hence an inconsistency. In the case of the inconsistency, there is usually a choice as to which constraint may be removed, since inconsistency is a property of a set of constraints and not of an individual constraint.

From the preceding discussion it can be seen that after a linear programming problem has been set up, we should proceed to solve it as if we expected to obtain an optimal solution. With this procedure, we either obtain the optimal solution or discover its shortcomings.

4-8 APPLICATION IN STEEL FABRICATING

Consider the following application in steel fabricating. A steel supplier buys stock directly from the steel mill, cuts to size, and sells to his customers. For one particular product, chrome bars 1-in in diameter he supplies any length up to 60-in. He himself buys the bars in 100-in lengths and cuts them to size. He presently has orders as follows:

Size, in	Quantity ordered
45	862
36	341
31	87
14	216

He wishes to cut enough to at least fill all the orders and do this by using the least possible number of 100-in bars.

The problem can be solved by finding all the possible cutting patterns and then finding how many times each of these patterns should be used. It is easily seen that two 45-in pieces can be cut with 10-in of trim loss, two 36-in pieces and two 14-in pieces can be cut with no trim loss, and so on. The following tableau shows all the possible cutting patterns, with their respective trim losses; x_i is the number of times that the ith pattern will be used:

c_j / Widths, in	1 x_1	1 x_2	1 x_3	1 x_4	1 x_5	1 x_6	1 x_7	1 x_8	1 x_9	1 x_{10}	1 x_{11}	1 x_{12}	Requirements
45	2	1	1	1	0	0	0	0	0	0	0	0	862
36	0	1	0	0	2	1	1	1	0	0	0	0	341
31	0	0	1	0	0	2	1	0	3	2	1	0	87
14	0	1	1	3	2	0	2	4	0	2	4	7	216
Trim loss, in	10	5	10	13	0	2	5	8	7	10	13	2	

If one designates the matrix enclosed in the rectangle as **A**, the requirements vector as **b**, and the solutions vector as **x**, one can designate the vector **c** as $(1, 1, \ldots, 1)$. The **c** vector shown is not obvious, but the problem calls for the minimization of the number of bars and not trim loss. With these definitions, the mathematical statement of the problem becomes:

Find

$$\mathbf{x} \geqq \mathbf{0}$$

such that

$$\mathbf{Ax} \geqq \mathbf{b}$$

and such that

$$Z = \mathbf{cx} \text{ is a } minimum$$

By designating the surplus variables as x_{13}, x_{14}, x_{15}, and x_{16} and the artificial variables as x_{17}, x_{18}, x_{19}, and x_{20} and using the two-phase method, one can enter variables x_1, x_5, and x_6 to eliminate three of the artificial variables and then surplus x_{16} to eliminate the remaining artificial variable, which then leaves you with the optimal tableau.

The results are as follows: with only three cutting patterns used:

Widths, in	Times used			Total produced	Optimal solution
	431	$148\frac{3}{4}$	$43\frac{1}{2}$		
45	2	0	0	862	$x_1 = 431$
36	0	2	1	341	$x_5 = 148\frac{3}{4}$
31	0	0	2	87	$x_6 = 43\frac{1}{2}$
14	0	2	0	$297\frac{1}{2}$	$x_{16} = 81\frac{1}{2}$
Trim loss, in	10	0	2		Bars used $= 623\frac{1}{4}$

This solution requires some interpretation because of the fractions. You cannot cut a pattern a fractional number of times. However, you can see that the requirements are met exactly for the first three widths, and that there is an excess of $81\frac{1}{2}$ of the 14-in size produced. The total number of bars used is $623\frac{1}{4}$. If one manipulates the solution a little, as shown in the following table, it is possible to produce the exact number of bars required without increasing the number of 100-in bars used.

Another solution (below) has the advantage that, with the same number of total bars used, the requirements are met exactly by inserting a fourth pattern, which has 28-in of trim loss, which is greater than the fourth size in the problem. That means that the 28-in piece is held in inventory

Another solution

Widths, in	Times used 431	108	$43\frac{1}{2}$	$40\frac{3}{4}$	Total produced
45	2	0	0	0	862
36	0	2	1	2	341
31	0	0	2	0	87
14	0	2	0	0	216
Trim loss, in	10	0	2	28	

without cutting it into two 14-in pieces—perhaps an advantage. It is interesting to convert this to an integer solution. The student should do this.

This is a rather crude variation of a very important problem in the paper industry. You can see how complicated it might get if there were 20 sizes required, a very large bar from which the orders are cut and, maybe a maximum number of cuts from a bar. This might result in as many as 5000 cutting patterns. This becomes very cumbersome, and a much more sophisticated method is developed. The complete tableau for the optimal solution of this problem is as shown on page 99.

EXERCISES

The following exercises consist of a mixture of formal problems and applications. If the formal problems are solved first, any mathematical model that can be set up is handled more easily. The applications extend the methods of linear programming to new areas.

1 Solve the following problem by the simplex method.
Find

$$x_1, x_2 \geqq 0$$

such that

$$x_1 + x_2 \leqq 10$$
$$2x_1 + x_2 \geqq 8$$
$$x_1 + 2x_2 \geqq 10$$

and such that

$$z = 60x_1 + 60x_2 \text{ is a } maximum$$

2 Solve the problem of Example 1-4 (p. 8), by the simplex method. The mathematical model is
Find

$$x_1, x_2 \geqq 0$$

Optimal tableau

c_j		-1	-1	-1	-1	-1	-1	-1	-1	-1	-1	-1	-1	0	0	0	0	
Variables		x_1	x_2	x_3	x_4	x_5	x_6	x_7	x_8	x_9	x_{10}	x_{11}	x_{12}	x_{13}	x_{14}	x_{15}	x_{16}	000
x_1	-1	1	$\frac{1}{2}$	$\frac{1}{2}$	$\frac{1}{2}$	0	0	0	0	0	0	0	0	$-\frac{1}{2}$	0	0	0	431
x_{16}	0	0	0	$-\frac{3}{2}$	-3	0	0	$-\frac{3}{2}$	-3	$-\frac{3}{2}$	-3	$-\frac{9}{2}$	-7	0	-1	$\frac{1}{2}$	1	$\frac{163}{2}$
x_6	-1	0	0	$\frac{1}{2}$	0	0	1	$\frac{1}{2}$	0	$\frac{3}{2}$	1	$\frac{1}{2}$	0	0	0	$-\frac{1}{2}$	0	$\frac{87}{2}$
x_5	-1	0	$\frac{1}{2}$	$-\frac{1}{4}$	0	1	0	$\frac{1}{4}$	$\frac{1}{2}$	$-\frac{3}{4}$	$-\frac{1}{2}$	$-\frac{1}{4}$	0	0	$-\frac{1}{2}$	$\frac{1}{4}$	0	$\frac{595}{4}$
$z_j - c_j$		0	0	$\frac{1}{4}$	$\frac{1}{2}$	0	0	$\frac{1}{4}$	$\frac{1}{2}$	$\frac{1}{4}$	$\frac{1}{2}$	$\frac{3}{4}$	0	$\frac{1}{2}$	$\frac{1}{2}$	$\frac{1}{4}$	0	$-623\frac{1}{4}$

such that

$$x_1 - 2x_2 \leqq 0$$

$$-x_1 + x_2 \leqq 4$$

$$x_1 \geqq 1$$

and such that

$$z = -2x_1 + x_2 \text{ is a } maximum$$

3 Solve the diet problem in Exercise 2, Chap. 1 (p. 11). The mathematical model of this minimization problem is:

Find

$$x_1, x_2 \geqq 0$$

such that

$$2x_1 + x_2 \geqq 200$$

$$x_1 + 4x_2 \geqq 160$$

and such that

$$Z = 10x_1 + 15x_2 \text{ is a } minimum$$

4 Solve the sweeper and polisher problem given in Exercise 4, Chap. 1 (p. 12), by the simplex method.

The solutions of parts (*a*) and (*b*) are straightforward applications of the simplex method to an interesting problem.

In solving parts (*c*) and (*d*), first obtain the solution using the regular two variables and try to answer the question of extra runs by noticing the effect of changing the right-hand side of the constraint initially restricted to 20 runs.

Second, solve parts (*c*) and (*d*) by letting x_3 = the number of additional runs to be rented at 25 cents each. This new variable will appear in the objective function and in the constraint on machine runs. As two parts of this problem, first restrict the number of extra runs to five and then remove this constraint entirely so that the optimal number of extra runs can be obtained under each of the conditions.

5 Set up the initial tableau for the shipping schedule problem given in Exercise 3, Chap. 2 (p. 56).

Perform one iteration of the simplex method. In the selection of the first variable to be entered, you will note that there is a tie in the $z_j - c_j$ row. How do you decide which variable is to be entered when there is a tie? Why? A tie occurs when the two-phase method is used.

Determine which variable should be entered at the second stage. Does it necessarily happen that an artificial variable will be removed at each iteration until all artificial variables are removed? Explain.

Perform another iteration. See if you can begin to make up an algorithm for this scheduling problem that could be explained to a person who does not know linear programming.

6 Consider the steel fabricating problem in Sec. 4-8. Suppose that the supplier has found that he seems to have too much waste when buying only 100-in lengths of the chrome bars. He decides that he should also stock 80-in lengths from which to cut his orders.

(*a*) For the mix of orders given on p. 96, what is the minimum number of 80-in bars that would be required to fill all the orders? In terms of the total amount of chrome used, is this a better or worse solution than that given in Sec. 4-8? Note that the cutting patterns from the 80-in bars are as follows:

45	1	1	0	0	0	0	0	0
36	0	0	2	1	1	0	0	0
31	1	0	0	1	0	2	1	0
14	0	2	0	0	3	1	3	5
Trim loss, in	4	7	8	13	2	4	7	10

(*b*) For the same mix of orders given on p. 96, and with a supply of both 80-in bars and 100-in bars, how should the orders be cut from bars of these two sizes in order to use the least amount of chrome metal from the stock supply, keeping in mind that when cutting from an 80-in bar, you are using only 80 percent as much chrome as when cutting from a 100-in bar? To do this, you must consider the 12 cutting patterns from the 100-in bars and the 8 cutting patterns from the 80-in bars. Also keep in mind that when you have a choice from the two sizes, the result must be at least as good as the equivalent of using $623\frac{1}{4}$ of the 100-in bars, since it would then call for no cuts to be made from the 80-in bars. If, however, some cuts are made from the 80-in bars, the number used would be converted to equivalent 100-in bars by multiplying the number of 80-in bars by 0.8, and this when added to the number of 100-in bars would give the total equivalent 100-in bars.

(*c*) Noting the solution on p. 97 to be $623\frac{1}{4}$ bars, it will obviously require 624 100-in bars. You will also note that two of the cutting patterns are used a fractional number of times, which really has no meaning since you either use the pattern or you do not. If you remove the $\frac{1}{2}$ from the $43\frac{1}{2}$ times used, and the $\frac{3}{4}$ from the $40\frac{3}{4}$, you reduce the number of bars used to 622, but you also find that you are two orders short on the 36-in size and one order short on the 31-in size. You obviously cannot cut these three sizes from one 100-in bar, but you can cut them from two bars, and therefore you can convert the optimal solution with fractional use to one with integer use by reducing the $43\frac{1}{2}$ to 43, increasing the $40\frac{3}{4}$ to 41, and cutting one 31-in size with a waste of 69-in all with 624 100-in bars.

CHAPTER

FIVE

SENSITIVITY ANALYSIS

5-1 PROBLEM MODIFICATION TO TEST SENSITIVITY

After a problem has been solved completely, it is often advantageous to investigate the effect on the solution of a change of some of the parameters. Such postoptimality analysis or sensitivity analysis is an important part of any discussion of linear programming. The following changes will be discussed in this chapter:

1. Changes in one or more of the components of the original **b** vector.
2. Changes in the coefficients of the objective function.
3. Addition of a new variable or activity vector to the optimal solution.
4. Changes in the coefficients of one or more of the constraints.
5. Addition of a new constraint to an optimal solution.
6. Removal of an existing constraint from an optimal solution.

5-2 DUALITY

To properly discuss the variations listed above, further background information must be developed. Returning now to the comments made in Sec. 2-3, let us pursue the concept of duality. As indicated at that time, corresponding to every linear programming problem stated in its original or primal form there is another linear programming problem, closely related to it, which we refer to as the *dual linear programming problem.*

As a formal matter, we should know how to write the mathematical statement of any pair of primal and dual problems. To do this, I shall arbitrarily assume that the primal problem has been written in the form of *maximization with* $\leqq$ *constraints.* This is no restriction, of course, since any problem can be written this way. The primal problem, then, as stated previously, is:

Find

$$\mathbf{x} \geqq \mathbf{0}$$

such that

$$\mathbf{Ax} \leqq \mathbf{b}$$

and such that

$$z = \mathbf{cx} \text{ is a } \textit{maximum}$$

With this assumption, the dual will always be:

Find

$$\mathbf{y} \geqq \mathbf{0}$$

such that

$$\mathbf{A'y} \geqq \mathbf{c'}$$

and such that

$$Z = \mathbf{b'y} \text{ is a } \textit{minimum}$$

You will note that in order for this dual statement to be an actual statement of a problem, the number of components in $\mathbf{y}$ must be equal to the number of constraints in the primal problem.

It is thus intuitively obvious that the components of $\mathbf{y}$, which are the solutions to the dual problem, have something to do with the constraints to the primal problem, since there is a one-to-one correspondence between them. This is, of course, true, as will be shown later. The relationship is that the solution component y_i is the marginal rate of change of the value of the objective function with respect to the change in the right-hand side b_i:

$$y_i = \lim_{\Delta b_i \to 0} \frac{\Delta z}{\Delta b_i} = \mathbf{c_B B^{-1}}$$

This relationship will be established later, and use of the property will be made in sensitivity analysis. For the problem stated in Chap. 2, pp. 13 and 14, we had

$$\mathbf{A} = \begin{bmatrix} 1.5 & 1 & 2.4 & 1 \\ 1 & 5 & 1 & 3.5 \\ 1.5 & 3 & 3.5 & 1 \end{bmatrix} \qquad \mathbf{A}' = \begin{bmatrix} 1.5 & 1 & 1.5 \\ 1 & 5 & 3 \\ 2.4 & 1 & 3.5 \\ 1 & 3.5 & 1 \end{bmatrix}$$

$$\mathbf{x} = \begin{bmatrix} x_1 \\ x_2 \\ x_3 \\ x_4 \end{bmatrix} \qquad \mathbf{b} = \begin{bmatrix} 2000 \\ 8000 \\ 5000 \end{bmatrix} \qquad \mathbf{y} = \begin{bmatrix} y_1 \\ y_2 \\ y_3 \end{bmatrix}$$

$$\mathbf{c} = (5.24, 7.30, 8.34, 4.18)$$

Readers should formulate this problem in both the primal and dual forms in order to solidify their understanding.

5-3 SOME FUNDAMENTAL PROPERTIES OF THE PRIMAL AND DUAL PROBLEMS

There are a number of interesting and important relationships between the primal and dual problems.

The inequality

$$\mathbf{Ax} \leqq \mathbf{b}$$

can also be written

$$\mathbf{x}'\mathbf{A}' \leqq \mathbf{b}' \tag{5-1}$$

Using any feasible $\mathbf{y}$ (meaning $\mathbf{y} \geqq \mathbf{0}$), we can multiply both sides of (5-1) on the right, to obtain

$$\mathbf{x}'\mathbf{A}'\mathbf{y} \leqq \mathbf{b}'\mathbf{y} = Z \tag{5-2}$$

Furthermore, considering the basic constraints for the dual problem, the inequality

$$\mathbf{A}'\mathbf{y} \geqq \mathbf{c}'$$

can also be written

$$\mathbf{c}' \leqq \mathbf{A}'\mathbf{y} \tag{5-3}$$

Using any feasible $\mathbf{x}$ (meaning $\mathbf{x} \geqq \mathbf{0}$), we can multiply both sides of (5-3) on the left to obtain

$$z = \mathbf{x}'\mathbf{c}' \leqq \mathbf{x}'\mathbf{A}'\mathbf{y} \tag{5-4}$$

Thus from (5-2) and (5-4) we can conclude that for any feasible solution $\mathbf{x}$ to the primal and any feasible solution $\mathbf{y}$ to the dual,

$$z = \mathbf{cx} = \mathbf{x}'\mathbf{c}' \leqq \mathbf{x}'\mathbf{A}'\mathbf{y} \leqq \mathbf{b}'\mathbf{y} = Z$$

That is, *for any feasible solutions to the primal and dual problems, the value of the primal objective function being maximized is always equal to or less than the value of the dual objective function being minimized.*

A second property of considerable importance is that if $\hat{\mathbf{x}}$ is a feasible solution to the primal and $\hat{\mathbf{y}}$ is a feasible solution to the dual such that $\mathbf{c}\hat{\mathbf{x}} = \mathbf{b}'\hat{\mathbf{y}}$, then $\hat{\mathbf{x}}$ is an optimal solution to the primal and $\hat{\mathbf{y}}$ is an optimal solution to the dual.

This can be shown as follows. For any feasible $\mathbf{x}$, and the corresponding particular feasible $\hat{\mathbf{y}}$, from the property just shown above we have

$$\mathbf{cx} \leqq \mathbf{b}'\hat{\mathbf{y}}$$

but since we have just assumed that

$$\mathbf{c}\hat{\mathbf{x}} = \mathbf{b}'\hat{\mathbf{y}}$$

by substitution, we have

$$\mathbf{cx} \leqq \mathbf{c}\hat{\mathbf{x}}$$

which is proof, of course, that $\mathbf{c}\hat{\mathbf{x}}$ is the maximum value.

Similarly, for any feasible $\mathbf{y}$ and corresponding feasible $\hat{\mathbf{x}}$,

$$\mathbf{c}\hat{\mathbf{x}} \leqq \mathbf{b}'\mathbf{y}$$

but since we assume that

$$\mathbf{c}\hat{\mathbf{x}} = \mathbf{b}'\hat{\mathbf{y}}$$

we have

$$\mathbf{b}'\hat{\mathbf{y}} \leqq \mathbf{b}'\mathbf{y}$$

indicating that $\mathbf{b}'\hat{\mathbf{y}}$ is the minimum value. *Thus if we are able to find a feasible* $\hat{\mathbf{x}}$ *and a feasible* $\hat{\mathbf{y}}$ *such that* $\mathbf{c}\hat{\mathbf{x}} = \mathbf{b}'\hat{\mathbf{y}}$, *we have optimal solutions to both the primal and dual problems.*

The third property that we wish to point out is that if either the primal or dual problem has an optimal solution, the other one does, too. The proof is made by actually constructing an optimal solution to the dual from an optimal solution to the primal.

The proof will be constructed from the primal statement of the problem in a form with equal-to-or-less-than constraints:

Find

$$\mathbf{x} \geqq \mathbf{0}$$

such that

$$\mathbf{Ax} \leqq \mathbf{b}$$

and such that

$$z = \mathbf{cx} \text{ is a } maximum$$

Since this statement is to represent all linear programming problems, we cannot guarantee that the components of the **b** vector are necessarily non-negative. If we now let $\mathbf{x}_s \geqq 0$ be the vector of slack variables added to convert the inequalities to equations, the problem can be stated:

Find

$$\mathbf{x} \geqq \mathbf{0} \qquad \mathbf{x}_s \geqq \mathbf{0}$$

such that

$$\mathbf{Ax} + \mathbf{Ix}_s = \mathbf{b}$$

and such that

$$z = \mathbf{cx} \text{ is a maximum}$$

For this primal problem, let $\mathbf{x_B}$ be an *optimal* basic feasible solution, **B** the corresponding basis matrix, and $\mathbf{c_B}$ the price vector of the basic variables. Then

$$z_j - c_j \geqq 0 \qquad \text{for all } j$$

and in particular for the vectors $\mathbf{a}_j$ of **A**. Therefore,

$$z_j - c_j = \mathbf{c_B}\mathbf{y}_j - c_j = \mathbf{c_B}\mathbf{B}^{-1}\mathbf{a}_j - c_j \geqq 0 \qquad \text{(for all } j)$$

which is equivalent to

$$\mathbf{c_B}\mathbf{B}^{-1}\mathbf{A} \geqq \mathbf{c} \tag{5-5}$$

when written in matrix form. If we now let

$$\mathbf{y}' = \mathbf{c_B}\mathbf{B}^{-1} \tag{5-6}$$

and substitute (5-6) into (5-5), we have

$$\mathbf{y}'\mathbf{A} \geqq \mathbf{c} \tag{5-7}$$

or, if written in transposed form,

$$\mathbf{A}'\mathbf{y} \geqq \mathbf{c}'$$

Thus we see that $\mathbf{y}' = \mathbf{c_B}\mathbf{B}^{-1}$ is a solution to the dual problem as written on p. 17. If we consider equation (5-5) for the portion corresponding to the slack variables, we have

$$\mathbf{c_B}\mathbf{B}^{-1}\mathbf{I} = \mathbf{c_B}\mathbf{B}^{-1} = \mathbf{y}' \geqq \mathbf{c_I} = \mathbf{0}$$

and hence the constructed solution is feasible. Furthermore,

$$Z = \mathbf{y}'\mathbf{b} = \mathbf{c_B}\,\mathbf{B}^{-1}\mathbf{b} = \mathbf{c_B}\,\mathbf{x_B} = \text{maximum } z$$

Hence we have shown that $\mathbf{x_B}$ and $\mathbf{c_B}\,\mathbf{B}^{-1}$ are feasible solutions to the primal and dual problems, respectively, for which $z = Z$; that is, the objective functions are equal, so $\mathbf{c_B}\,\mathbf{B}^{-1}$ is an optimal solution to the dual problem.

In summary: *If there is an optimal solution to the primal problem, we have proved by construction that there is an optimal solution to the dual problem. In fact, if* **B** *is the basis matrix for the primal problem corresponding to an optimal solution and if* $\mathbf{c_B}$ *contains the prices in the basis, then an optimal solution to the dual is*

$$\mathbf{y}' = \mathbf{c_B}\,\mathbf{B}^{-1}$$

5-4 COMPUTATIONS

To apply this to the example problem that we have been using, refer to the summary solution on p. 70. For the optimal tableau, we see that

$$\mathbf{c_B} = (c_1, c_4, c_2) = (5.24, 4.18, 7.30)$$

and

$$\mathbf{B}^{-1} = \begin{bmatrix} 0.6471 & -0.2353 & 0.1765 \\ 0.5294 & 0.3529 & -0.7647 \\ -0.5000 & 0.0000 & 0.5000 \end{bmatrix}$$

Consequently,

$$\mathbf{c_B}\,\mathbf{B}^{-1} = (1.9537, 0.2424, 1.3782)$$

to within rounding error. These values are seen to be located in the final tableau in the $z_j - c_j$ row in the columns corresponding to the slack variables. This, of course, assists us in the interpretation of these dual solutions. Since the first component above is in the $z_j - c_j$ row under the slack variable corresponding to the first constraint, which has a right-hand side of 2000, it tells us that if we were to enter a unit of slack in this department, the penalty would be 1.9537. That is equivalent to saying that if we lowered the right-hand-side element of 2000 by 1 unit, the profit would be reduced by \$1.9537. Suppose that we reverse this situation and, instead of lowering the right-hand side, raise the right-hand side by 1 unit. This is then in the opposite direction, and therefore the profit would be increased and the per unit rate of increase would be \$1.9537. Thus we can say that the first component of the vector $\mathbf{c_B}\,\mathbf{B}^{-1}$ represents the marginal value of 1 unit of resource of the right-hand side of the first constraint.

At this point no indication is made as to how long that rate of improvement of profit would hold if one were to continue to increase the right-hand side above the original value of 2000. This will be taken care of very shortly, however. The second component of 0.2424 gives the marginal value of a unit of resource of the right-hand side of the second constraint, and similarly the third component of 1.3782 is the marginal value of a unit of resource of the right-hand side of the third constraint.

5-5 SIMPLEX SOLUTION FOR GENERAL PROBLEMS

In order to assist with sensitivity analysis, let us clarify some points about the simplex method of solution. In order to be systematic, most people will search through the constraints and add the negative of the surplus variables first. After this, the variables supplying the unit vectors are added. These supply the slack variables for the $\leq$ constraints and the artificial variables for the remainder of the constraints. The net result is that the initial tableau has the following schematic arrangement:

Original variables of the problem	Surplus variables	Slack variables	Artificial variables for the $\geq$ constraints	Artificial variables for the $=$ constraints
For the $\leq$ constraints	0	I_p	0	0
For the $\geq$ constraints	-1's	0	I_q	0
For the $=$ constraints	0	0	0	I_r
Final tableau $(z_j - c_j)$'s	$-\mathbf{c_B}\,\mathbf{B}^{-1}$	$\mathbf{c_B}\,\mathbf{B}^{-1}$	$\mathbf{c_B}\,\mathbf{B}^{-1} + M$	$\mathbf{c_B}\,\mathbf{B}^{-1} + M$

This format has the advantage that the identity matrix in the initial tableau appears at the extreme right of the tableau. As a result of this arrangement, $\mathbf{B}^{-1}$ will appear on the extreme right of the optimal tableau.

5-6 DUAL VALUES CORRESPONDING TO EQUAL-TO-OR-LESS-THAN CONSTRAINTS

As indicated, the dual solutions corresponding to equal-to-or-less-than constraints are found in the final tableau in the $z_i - c_i$ row in the columns corresponding to the respective slack variables. This is to be expected, since

$$z_j - c_j = \mathbf{c_B}\,\mathbf{B}^{-1}\mathbf{e}_j - c_j = \mathbf{c_B}\,\mathbf{B}^{-1}$$

This results from the fact that $c_j = 0$ whenever j refers to a slack variable, and correspondingly the $\mathbf{e}_j$ (p. 22) are unit vectors. Thus if a problem has all equal-to-or-less-than constraints and the components of the right-hand side are all initially nonnegative, the primal solution can be obtained, the $\mathbf{c_B}$ vector and $\mathbf{B}^{-1}$ matrix can be easily located, and so the dual solutions can be computed immediately. *Note that for this case* $\mathbf{c_B B}^{-1}$ *must necessarily be either zero or positive.* It is important to remember this fact for future analysis.

5-7 DUAL VALUES CORRESPONDING TO EQUAL-TO-OR-GREATER-THAN CONSTRAINTS

Consider now what happens when the primal problem has an $\geqq$ constraint with the right-hand side being positive. To consider it in the framework of the previous analysis, the constraint would have to be written as an $\leqq$ constraint, with the right-hand side being negative. In this format there would be a positive slack variable and $\mathbf{c_B B}^{-1}$ would be a nonnegative quantity. The simplex method of solution would use the $\geqq$ constraint to keep the right-hand side positive and to maintain feasible solutions at each iteration. Thus when evaluating the $\mathbf{c_B B}^{-1}$ component for the dual solution, it would be found that the value corresponding to an $\geqq$ constraint would be nonpositive, since in the actual solution, the column in $\mathbf{B}^{-1}$ that relates to this constraint comes from the artificial variable, which has the opposite sign from the surplus variable. Therefore, to find the actual dual variable, the sign would have to be reversed. It will be important to *note, then, that corresponding to an* $\geqq$ *constraint,* $\mathbf{c_B B}^{-1}$ *would be either negative or zero* when the problem is solved by the regular simplex method. The negative sign will have important implications, as will be shown later.

5-8 DUAL VALUES CORRESPONDING TO EQUAL-TO CONSTRAINTS

Consider now what happens when we have an equal-to constraint. We can always write this constraint so that the right-hand side will be positive or zero. To consider it in the framework of the preceding analysis, we must first write the constraint as two constraints, one $\geq$ and the other $\leq$. If we now apply the computation of $\mathbf{c_B B}^{-1}$ to each of these, we would find in the one case that the result would be nonpositive and in the other case that it would be nonnegative. Since the regular simplex method of solution does not require that you separate the constraint into two constraints, the *computation of* $\mathbf{c_B B}^{-1}$ *may lead to either a positive or a negative or a zero.* The dual solution will, of course, be the absolute value of the result. If the result is positive, it means that the $\leqq$ part of the constraint is binding and if the result is negative, the $\geqq$ part of the constraint is binding.

5-9 SUMMARY OF DUAL SOLUTIONS

Although the analysis above assumes that the primal problem has been written as a maximization problem with $\leqq$ constraints, it is not necessary to do so to find the dual solutions. The problem should be solved in the normal fashion with the **b** vector having nonnegative components. Then by computing $\mathbf{c_B B}^{-1}$ and taking absolute values, the dual solutions are obtained. The solutions are in order of the constraints. Any positive value indicates that the gains in the objective function are to be made by *raising* the right-hand side. Any negative value indicates that the gains in the objective function are to be made by *lowering* the right-hand side. In either case the gain is to be made by *relaxing* the constraint. Thus a positive value could only be obtained from an equal-to-or-less-than constraint or an equal-to constraint acting like an equal-to-or-less-than constraint. A negative value could only be obtained from an equal-to-or-greater-than constraint or an equal-to constraint acting like an equal-to-or-greater-than constraint. A zero value could actually be obtained from any one of the three types of constraints. The preceding facts will be especially important in ranging analysis.

5-10 PROCESS FOR WRITING THE DUAL STATEMENT OF A PROBLEM

To get a deeper insight into the relationship between a primal problem and its dual, consider the problem stated on p. 85 in the standard primal form. To write the dual statement of the problem it is necessary first to write it with all $\leqq$ constraints. This results in the following form of the primal statement:

Find

$$x_1, x_2, x_3 \geqq 0$$

such that

$$\begin{aligned} x_1 + x_2 + x_3 &\leqq 500 \\ -x_1 - x_2 \qquad &\leqq -100 \\ x_1 - x_2 - x_3 &\leqq 120 \\ -x_1 + x_2 + x_3 &\leqq -120 \end{aligned} \tag{5-8}$$

and such that

$$z = 5x_1 + 10x_2 + 15x_3 \text{ is a } \textit{maximum}$$

In this form the matrix of coefficients is

$$\mathbf{A} = \begin{bmatrix} 1 & 1 & 1 \\ -1 & -1 & 0 \\ 1 & -1 & -1 \\ -1 & 1 & 1 \end{bmatrix}$$

and therefore

$$\mathbf{A}' = \begin{bmatrix} 1 & -1 & 1 & -1 \\ 1 & -1 & -1 & 1 \\ 1 & 0 & -1 & 1 \end{bmatrix}$$

Consequently, the dual statement will be:

Find

$$y_1, y_2, y_3, y_4 \geqq 0$$

such that

$$\begin{aligned} y_1 - y_2 + y_3 - y_4 &\geqq 5 \\ y_1 - y_2 - y_3 + y_4 &\geqq 10 \\ y_1 \qquad - y_3 + y_4 &\geqq 15 \end{aligned} \tag{5-9}$$

and such that

$$Z = 500y_1 - 100y_2 + 120y_3 - 120y_4 \text{ is a } \textit{minimum}$$

If one solves this dual problem, he will get the same answer as is obtained by computing

$$\mathbf{c_B B}^{-1} = (15, 5, 0) \begin{bmatrix} 0.5 & 0 & -0.5 \\ 0.5 & 0 & 0.5 \\ 0.5 & -1 & 0.5 \end{bmatrix} = (10, 0, -5)$$

adjusted for whether the constraint is $\leqq$, $\geqq$, or $=$. The first component in the solution comes from an $\leqq$ constraint, and hence the result remains as is (i.e., $y_1 = 10$). The second component comes from an $\geqq$ constraint, and hence the sign should be reversed; but since it is zero, it makes no difference and $y_2 = 0$. The third component comes from the combination $y_3 - y_4$, where each of the y_3 and y_4 are $\geqq 0$, but the combination is unrestricted in sign. In this case it leads to $y_3 = 0$ and $y_4 = 5$ and to $y_3 - y_4 = -5$, just as shown in the computation. If you examine the dual statement as shown in (5-9), you will notice that the variables y_3 and y_4 exist in the combination $y_3 - y_4$ throughout the problem, and therefore we could replace the combination by a single variable y_5, which would be unrestricted in sign, since we do not know at the outset whether y_3 or y_4 takes on the larger value.

Incorporating this idea, a new statement of the problem (5-9) would be:

Find

$$y_1, y_2 \geqq 0 \text{ and } y_5 \text{ (unrestricted in sign)}$$

such that

$$\begin{aligned} y_1 - y_2 + y_5 &\geqq 5 \\ y_1 - y_2 - y_5 &\geqq 10 \\ y_1 \qquad - y_5 &\geqq 15 \end{aligned} \tag{5-10}$$

and such that

$$Z = 500y_1 - 100y_2 + 120y_5 \text{ is a } \textit{minimum}$$

In this statement of the problem, the one variable y_5 corresponds to the equal-to constraint, whereas we used two nonnegative variables originally to represent the equal-to constraint. In using $\mathbf{c_B B^{-1}}$ to find the dual solutions, the single value y_5 is found rather than each of y_3 and y_4. This is not only simpler, but it also has the advantage that we can tell by the sign of the result whether the $\geqq$ or the $\leqq$ portion of the constraint is being applied.

As stated before, solving either the primal or dual problem gives all the information about the solution. However, often there will be considerable difference in the time required for the solution of these two problems.

5-11 CHANGES IN THE COMPONENTS OF THE RIGHT-HAND SIDE

Returning to the problems of post optimality analysis, consider the effects of changing one or more of the components of the **b** vector. In connection with our example in Chap. 3, it was shown on p. 80 that the optimal solution was given by

$$\mathbf{x_B} = \mathbf{B}^{-1}\mathbf{b} = \begin{bmatrix} 294.11 \\ 58.82 \\ 1500.00 \end{bmatrix}$$

As indicated on p. 107, the dual solution corresponding to the first constraint is 1.9537 (i.e., the marginal value of an additional hour in machine center A is \$1.9537). We cannot tell from the dual solution itself the number of units for which this rate of increase of profit will hold—it is an instantaneous rate with a specific range of values over which it will hold.

By making use of the inverse of the basis matrix, one can evaluate these limits. Let θ be the number of units beyond 2000 for which this rate holds, so that

$$\begin{bmatrix} x_1 \\ x_4 \\ x_2 \end{bmatrix} = \overset{\mathbf{B}^{-1}}{\begin{bmatrix} 0.6471 & -0.2353 & 0.1765 \\ .5294 & 0.3529 & -0.7647 \\ -0.5000 & 0.0000 & 0.5000 \end{bmatrix}} \overset{\mathbf{b}}{\begin{bmatrix} 2000 + \theta \\ 8000 \\ 5000 \end{bmatrix}} = \overset{\mathbf{x_B}}{\begin{bmatrix} 294.11 + 0.6471\theta \\ 58.82 + 0.5294\theta \\ 1500.00 - 0.5000\theta \end{bmatrix}} \quad (5\text{-}11)$$

The problem of evaluating the limits is one of finding how much θ can change while maintaining the same basic variables in the solution, so it will be necessary that

$$\begin{aligned} 294.11 + 0.6471\theta &\geqq 0 \\ 58.82 + 0.5294\theta &\geqq 0 \\ 1500.00 - 0.5000\theta &\geqq 0 \end{aligned} \quad (5\text{-}12)$$

The first two of these inequalities place no limitation on the increase in the number of hours in machine center A. The third relationship requires that

$$\begin{aligned} 1500.00 - 0.5000\theta &\geqq 0 \\ 0.5000\theta &\leqq 1500.00 \\ \theta &\leqq 3000 \end{aligned}$$

The interpretation is that if the hours in machine center A are increased by 3000, to a new value of 5000, then 3000(1.9537) = \$5861.10 of additional profits are obtained. If the additional hours can be made available for less than this amount, a net profit is realized. Looking at the complete solution with $\theta = 3000$, we find that

$$\begin{aligned} x_1 &= 294.11 + 0.6471(3000) = 2235.41 \\ x_4 &= 58.82 + 0.5294(3000) = 1647.02 \\ x_2 &= 1500.00 - 0.5000(3000) = 000 \end{aligned}$$

(i.e., we have a degenerate solution). If we were to increase the hours by more than 3000, the value of x_2 would become nonfeasible, and then this nonfeasibility would have to be eliminated. This would lead to a different set of basic variables in the program. The net effect would be that even though there might be additional profits, the rate of increase would no longer be 1.9537. The method of eliminating the nonfeasibility is the dual-simplex method, which will be discussed later.

Similar analyses can be made for the hourly restrictions on machine centers *B* and *C*. The optimal tableau shows that if the hours are increased in machine center *B*, the per hour profit is \$0.2424 and for machine center *C*, the rate is \$1.3782. The upper limits on these increases at the stated rates are separate calculations of the same type as above. The upper limit for machine center *B* is 1249.98 and the upper limit for machine center *C* is 76.92.

Returning to the right-hand side of the first constraint, and looking at inequalities (5-12), it is easily seen that θ can be negative (i.e., the right-hand side can be reduced), in which case the profits will be reduced. To find the lower limit, we must find the more restricting of the first two constraints in (5-12). Solving these two, we find: from

$$294.11 + 0.6471\theta \geqq 0$$

$$0.6471\theta \geqq -294.11$$

$$\theta \geqq -454.50$$

and from

$$58.82 + 0.5294\theta \geqq 0$$

$$0.5294\theta \geqq -58.82$$

$$\theta \geqq -111.11$$

Since the latter is more restrictive, we combine it with the result for the upper limit and write

$$-111.11 \leqq \theta \leqq 3000$$

That is, the right-hand side can range from

$$2000 - 111.11 \quad \text{to} \quad 2000 + 3000$$

or from

$$1888.89 \quad \text{to} \quad 5000$$

without changing the basic variables. The per unit rate of change of profit within that range is \$1.9537, and increases when the right-hand side is increased and decreases when the right-hand side is decreased.

We might appropriately arrange these results as follows:

Min. b_1 z lower	Original b_1 z	Max. b_1 z upper
1888.89 12,520	2,000 12,737	5,000 18,598

For the preceding analysis to hold, all other parameters of the problem must remain fixed. Similar results for b_2 are shown in the following table:

Min. b_2 z lower	Original b_2 z	Max. b_2 z upper
7,833.3 12,697	8,000 12,737	9,250.0 13,040

The results for the right-hand side of the third constraint are:

Min. b_3 z lower	Original b_3 z	Max. b_3 z upper
3,333.3 10,440	5,000 12,737	5,076.9 12,843

It must be emphasized that the results above must be considered independently.

If one wishes to change more than one of the right-hand sides, one method of handling the effects is to apply the preceding methods sequentially. For example, suppose that the right-hand side b_1 is increased up to 3000 and then you would like to find both the effects and the range for changing the right-hand side b_2.

Considering the change of b_1 from 2000 to 3000, the objective function will increase by an amount 1000 times 1.9537, yielding an increase of \$1953.7. From (5-11) we see that

$$\begin{aligned} x_1 &= 294.11 + 647.10 = 941.21 \\ x_4 &= 58.82 + 529.40 = 588.22 \\ x_2 &= 1500.00 - 500.00 = 1000.00 \end{aligned}$$

If we now wish to evaluate the effects of changing b_2, we must study the effects of changes θ in the equations

$$\begin{bmatrix} x_1 \\ x_4 \\ x_2 \end{bmatrix} = \begin{bmatrix} 0.6471 & -0.2353 & 0.1765 \\ 0.5294 & 0.3529 & -0.7647 \\ -0.5000 & 0.0000 & 0.5000 \end{bmatrix} \begin{bmatrix} 3000 \\ 8000 + \theta \\ 5000 \end{bmatrix}$$

$$= \begin{bmatrix} 941.21 - 0.2353\theta \\ 588.22 + 0.3529\theta \\ 1000.00 + 0.0000\theta \end{bmatrix}$$

from which it is seen that θ can increase 4000 and can decrease 1666.82 while the objective function changes at the rate of 0.2424 as provided by the dual solution. Thus the range for the right-hand side is from 6333.18 to 12,000 rather than from 7833.33 to 9250, as computed before the right-hand side of the first constraint was raised from 2000 to 3000.

The right-hand portion of Eqs. (5-11) show the new solution values due to changes, θ, in the right-hand side of the first constraint. To consider the effects of right-side values intermediate to the range, $\mathbf{B}^{-1}$ and $\mathbf{x_B}$ from the optimal tableau are recorded:

$\mathbf{x_B}$	$\mathbf{c'_B}$		$\mathbf{B}^{-1}$		$\mathbf{x_B}$
x_1	5.24	0.6471	−0.2353	0.1765	294.11
x_4	4.18	0.5294	0.3529	−0.7647	58.82
x_2	7.30	−0.5000	0.0000	0.5000	1500.00

These are the necessary items of information for finding intermediate solutions, as well as the limiting values for changes in the right-hand side of each of the constraints.

The left-hand column of $\mathbf{B}^{-1}$ is the vector in the final tableau which has been updated from x_5, the slack in machine center A. These elements provide the rates of change of the solutions x_1, x_4, and x_2 for a unit increase in the right-hand side of the first constraint. These rates of change would be reversed if the right-hand side were decreased. Thus, if the 2000 in the first constraint were increased by 1 unit, x_1 is increased by an amount 0.6471 from its value of 294.11; x_4 is increased by an amount of 0.5294 from its value of 58.82; and x_2 is *decreased* by an amount of 0.5000 from its value of 1500. The amount of the increase at this rate is therefore limited by the number of hours required to reduce the 1500 to zero. The limit is 1500 divided by 0.5000.

The second column of $\mathbf{B}^{-1}$ provides the rates of change of the solutions x_1, x_4, and x_2 for a unit increase in the right-hand side of the second constraint. Thus x_1 is reduced by 0.2353, x_4 is increased by 0.3529, and x_2 remains unaltered. The upper limit for which this holds true is 294.11 divided by 0.2353, and the lower limit is 58.82 divided by 0.3529.

The third column of $\mathbf{B}^{-1}$ provides the rates of change of the solutions x_1, x_4, and x_2 for a unit increase in the right-hand side of the third constraint. Thus x_1 is increased 0.1765 unit, x_4 is decreased 0.7647 unit, and x_2 is increased by 0.5000 unit. The upper limit is 58.82 divided by 0.7647, and the lower limit is 294.11 divided by 0.1765.

The discussion above refers to maximization problems in which all the constraints are of the $\leqq$ variety. The methods apply equally well regardless of the nature of the constraints. There are several things that need be noted carefully.

1. The per unit rate of change of the objective function per unit change in the right-hand side of a constraint is given by $\mathbf{c_B B^{-1}}$.
2. For maximization problems, the algebraic sign of the component of $\mathbf{c_B B^{-1}}$ will indicate the direction in which the particular right-hand side should be moved to increase the objective function, a positive indicating that it should be increased, a negative indicating that it should be decreased, and a zero indicating that movement does not change the objective function. In the latter case, the constraint is nonbinding.
3. The columns of the basis inverse $\mathbf{B^{-1}}$ will give the rates of change of the solution components of $\mathbf{x_B}$ for an increase in the corresponding component of the right-hand side, the first column of $\mathbf{B^{-1}}$ referring to the first right-hand-side component, the second to the second, and so on.
4. Minimization problems are handled by conversion to maximization problems and translating the effect.

5-12 APPLICATION

Let us now apply the principles to the example on p. 85, whose optimal solution appears on p. 88, with the dual solution on p. 111 as computed from the primal problem:

$$\mathbf{c_B B^{-1}} = (15, 5, 0)\begin{bmatrix} 0.5 & 0 & -0.5 \\ 0.5 & 0 & 0.5 \\ 0.5 & -1 & 0.5 \end{bmatrix} = (10, 0, -5)$$

The first component, which is 10, indicates that profits will be increased at the marginal rate of 10 per unit change in the right-hand-side term of 500. There is no upper limit on this increase, since there are no negative terms in the first column of $\mathbf{B^{-1}}$.

The $y_2 = 0$ indicates that in the second original constraint, the 100 is no limitation at all. Lowering or relaxing it will not help. This is also easily seen by a simple analysis, since the optimal solution yields $x_1 + x_2 = 310$, which is 210 units beyond that given in the constraint.

The $y_3 = -5$ indicates that the third original constraint is operating as an equal-to-or-greater-than constraint and that gains would be made by lowering the value of 120, with the rate of increase of profit being \$5 per unit lowered. Lowering the value to 20, for instance, would increase the profits by \$500. Analysis shows that the value could be lowered by a total of 420 from the original, that is, lowering it from the original 120 to a value of -300, or making the constraint read

$$x_1 - x_2 - x_3 = -300$$

which is equivalent to

$$-x_1 + x_2 + x_3 = 300$$

5-13 CHANGES IN THE COEFFICIENTS OF THE OBJECTIVE FUNCTION

The second problem of postoptimality analysis deals with the important effects of changes in the solution which result from changes in the coefficients of the objective function. These changes must be divided into two categories: the c_j's corresponding to *nonbasic* variables and those corresponding to *basic* variables. For the case of a single *nonbasic* variable, one needs only substitute the new value of c_j into the corresponding $z_j - c_j$ and then see if it still tests optimal (i.e., is $z_j - c_j \geqq 0$?). For this category, the z_j is not changed, and therefore the test is simple. If the c_j has been increased enough to turn $z_j - c_j$ negative, the new solution is no longer optimal, and the jth variable will be entered into the solution.

In checking the optimal solution for the problem on p. 69, consider the nonbasic variable x_3. If c_3 is increased by more than 1.4146, then $z_3 - c_3$ will become negative and x_3 must be brought into the solution. There is no guarantee that one iteration will achieve the optimum, but x_3 will have to remain in the optimal solution.

For the case of a single *basic* variable, one again needs to examine the test for optimality, which for this case means to examine the entire $z_j - c_j$ row. The reason for this is that since we are considering the case for a basic variable, every z_j is affected, since $z_j = \mathbf{c_B} \mathbf{B}^{-1}\mathbf{a}_j$ and one of the components of $\mathbf{c_B}$ has been changed. Thus the entire $z_j - c_j$ row must be recomputed. It is, of course, conceivable that with the new value of $\mathbf{c_B}$, none of the $(z_j - c_j)$'s have turned negative, in which case the solution remains optimal, with only the possibility that the value of the objective function has changed. On the other hand, several of these values may have turned negative, in which case one continues with the solution in the normal fashion. Again consider our example and examine the range over which $c_1 = 5.24$ can change without changing the basis set of variables. To do this, let $c_1 = 5.24 + \theta$ in the optimal tableau on p. 69, so that we have:

Optimal tableau

	c_j	x_1 $5.24 + \theta$	x_2 7.30	x_3 8.34	x_4 4.18	x_5 0.00	x_6 0.00	x_7 0.00	b 000
x_1	$5.24 + \theta$	1.0000	0.0000	1.9353	0.0000	0.6471	−0.2353	0.1765	294.11
x_4	4.18	0.0000	0.0000	−1.0529	1.0000	0.5294	0.3529	−0.7647	58.82
x_2	7.30	0.0000	1.0000	0.5500	0.0000	−0.5000	0.0000	0.5000	1500.00
$z_j - c_j$		0.0000	0.0000	1.4146 $+1.9353\theta$	0.0000	1.9537 $+0.6471\theta$	0.2424 -0.2353θ	1.3782 $+0.1765\theta$	12,737 $+294.11\theta$

From the $z_j - c_j$ row, we now have:

$$z_1 - c_1 = z_2 - c_2 = z_4 - c_4 = 0$$

$$z_3 - c_3 = 1.4146 + 1.9353\theta \geqq 0$$

$$z_5 - c_5 = 1.9537 + 0.6471\theta \geqq 0$$

$$z_6 - c_6 = 0.2424 - 0.2353\theta \geqq 0$$

$$z_7 - c_7 = 1.3782 + 0.1765\theta \geqq 0$$

From $z_6 - c_6$, we see that

$$0.2353\theta \leqq 0.2424$$

$$\theta \leqq 1.0302$$

From $z_3 - c_3$, we see that

$$1.9353\theta \geqq -1.4146$$

$$\theta \geqq -0.7309$$

From $z_5 - c_5$, we see that

$$0.6471\theta \geqq -1.9537$$

$$\theta \geqq -3.0192$$

From $z_7 - c_7$, we see that

$$0.1765\theta \geqq -1.3782$$

$$\theta \geqq -7.8085$$

Of the last three constraints, we see that

$$\theta \geqq -0.7309$$

is the most restrictive, and combining it with the constraint obtained from $z_6 - c_6$, we have

$$-0.7309 \leqq \theta \leqq 1.0302$$

For this range of change in c_1 there will be no change in the basic variables in the optimal solution, although the value of the objective function will change by an amount of 294.11θ as long as θ stays in the restricted range listed above. This is equivalent to

$$5.24 - 0.73 \leqq c_1 \leqq 5.24 + 1.03$$

or

$$4.51 \leqq c_1 \leqq 6.27$$

The actual solution values remain the same. This means that the decision as to what mix of these products to produce will be the same, even though the profit will be slightly different.

The principles above can be applied to more than one c_j simultaneously, but when many changes are to be made, there is considerable doubt as to whether there is any saving in either time or computation by trying to adjust the optimal solution to meet the new conditions. It might well be better in such a case to solve the problem completely from the beginning with the new values of the c_j's inserted.

5-14 ADDITION OF A NEW VARIABLE

The third item of postoptimality analysis is that of adding a new variable with its corresponding activity vector. For our first example in this chapter, for instance, the question might arise as to whether an additional product x_p should be entered into the solution if it is known that the product requires 2 h in machine center A, 2 h in machine center B, and 1 h in machine center C, with a profit of \$6.38 per unit manufactured. It is quite obvious that the problem can be solved from the beginning with this new variable as one of the original variables, but it may often be easier to consider the effect of the new variable on the optimal solution. To do this we first find the updated vector corresponding to x_p by the transformation $\mathbf{B}^{-1}\mathbf{a}_p$, which is

$$\begin{bmatrix} 0.6471 & -0.2353 & 0.1765 \\ 0.5294 & 0.3529 & -0.7647 \\ -0.5000 & 0.0000 & 0.5000 \end{bmatrix} \begin{bmatrix} 2 \\ 2 \\ 1 \end{bmatrix} = \begin{bmatrix} 1.00 \\ 1.00 \\ -0.50 \end{bmatrix} = \mathbf{y}_p$$

That is, this is the vector that we would have in the final tableau had we entered the new variable initially and entered the other vectors in exactly the same order. It then remains to find out whether this vector should now be entered. Hence we compute

$$z_p - c_p = \mathbf{c_B}\,\mathbf{y}_p - c_p = (5.24, 4.18, 7.30) \begin{bmatrix} 1.00 \\ 1.00 \\ -0.50 \end{bmatrix} - 6.38$$

which yields -0.61. This indicates that the manufacture of the new product would yield additional profits over that of the original optimal solution. Since $z_j - c_j$ for this product indicates an improvement, one would perform another iteration by entering the variable x_p and continue until an optimal solution is reached. Had the per unit profit been \$5.77 or less, no gain could be attained and the original optimal solution would be retained. Several new variables could be tested in this way after having obtained an optimal solution for some lesser number of variables.

5-15 CHANGES IN THE ELEMENTS OF MATRIX A

The fourth item of postoptimality analysis deals with changes in one or more of the coefficients or components of one of the original activity vectors. These vectors are the columns of the **A** matrix. This happens to be a very complex problem, and as such can only be handled in part. It is important to recognize, however, those situations which can be handled with some ease and satisfaction. For this analysis we divide the activity vectors of **A** into two classes—those corresponding to *nonbasic* vectors and those corresponding to the *basic* vectors (i.e., those vectors that are in the solution).

Any nonbasic vector is handled in exactly the same way as is the addition of an entirely new activity vector. The original vector is removed, and the new vector is treated as an additional activity vector.

In the case of a basic vector, more complications arise. Consider, for instance, in our example problem that the vector $\mathbf{a}_4$ should have been

$$\begin{bmatrix} 3.5 \\ 1 \\ 1 \end{bmatrix} \text{ instead of } \begin{bmatrix} 1 \\ 3.5 \\ 1 \end{bmatrix}$$

as was used in the problem. We might consider that this had arisen as a typing or keypunching error, or just plain error in measurement or collection of data. The first step is to consider the vector as if it were a completely new activity vector and calculate its updated form by computing $\mathbf{B}^{-1}$ times this new vector:

$$\begin{bmatrix} 0.6471 & -0.2353 & 0.1765 \\ 0.5294 & 0.3529 & -0.7647 \\ -0.5000 & 0.0000 & 0.5000 \end{bmatrix} \begin{bmatrix} 3.5 \\ 1 \\ 1 \end{bmatrix} = \begin{bmatrix} 2.2061 \\ 1.4411 \\ -1.2500 \end{bmatrix}$$

Suppose that we now substitute this vector in place of the original updated $\mathbf{a}_4$ in the optimal tableau, to obtain:

	c_j	5.24	7.30	8.34	4.18	0.00	0.00	0.00	0000
Variables		x_1	x_2	x_3	x_4	x_5	x_6	x_7	b
$\mathbf{x_B}$	$\mathbf{c'_B}$								
x_1	5.24	1.0000	0.0000	1.9353	2.2061	0.6471	−0.2353	0.1765	294.11
x_4	4.18	0.0000	0.0000	−1.0529	1.4411	0.5294	0.3529	−0.7647	58.82
x_2	7.30	0.0000	1.0000	0.5500	−1.2500	−0.5000	0.0000	0.5000	1500.00
$z_j - c_j$		0.0000	0.0000	–	—	—	—	—	—

Now that we have replaced the $\mathbf{a}_4$ vector with an updated vector, you will note that it is not easy to read a basic feasible solution, since there are only two unit vectors instead of the three needed for a tableau with three constraints. If x_4 is to remain as a basis variable, its vector will have to be transformed into the unit vector

$$\begin{bmatrix} 0 \\ 1 \\ 0 \end{bmatrix}$$

so that x_1 and x_2 will remain basic variables. You will note that the $z_j - c_j$ row has not been computed. The reason for this is that the tableau does not have three unit vectors and must be transformed so that these are present.

Since the vector $\mathbf{y}_4$ must have a 1 in its second component, we shall divide the second row by 1.4411. After that, -2.2061 times this row will be added to the first row and 1.2500 times it will be added to the third row. With these new values, the $z_j - c_j$ row will be computed from $z_j - c_j = \mathbf{c_B}\,\mathbf{y}_j - c_j$ with the resulting tableau:

c_j		5.24	7.30	8.34	4.18	0.00	0.00	0.00	000
Variables		x_1	x_2	x_3	x_4	x_5	x_6	x_7	b
$\mathbf{x_B}$	$\mathbf{c'_B}$								
x_1	5.24	1.0000	0.0000	3.5471	0.0000	−0.1633	−0.7755	1.3471	204.07
x_4	4.18	0.0000	0.0000	−0.7306	1.0000	0.3674	0.2449	−0.5306	40.82
x_2	7.30	0.0000	1.0000	−0.3633	0.0000	−0.0408	0.3061	−0.1633	1,551.02
$z_j - c_j$		0.0000	0.0000	4.5408	0.0000	0.3822	−0.8056	3.6488	12,562

We now have a basic feasible solution with x_1, x_4, and x_2 as the basic variables. It is not optimal, however, since there is a negative in the $z_j - c_j$ row. x_6 is entered into the solution with x_4 being removed to obtain the optimal:

c_j		5.24	7.30	8.34	4.18	0.00	0.00	0.00	000
Variables		x_1	x_2	x_3	x_4	x_5	x_6	x_7	b
$\mathbf{x_B}$	$\mathbf{c'_B}$								
x_1	5.24	1.0000	0.0000	1.2333	3.1670	1.0000	0.0000	−0.3334	333.33
x_6	0.00	0.0000	0.0000	−2.9836	4.0836	1.5000	1.0000	−2.1669	166.67
x_2	7.30	0.0000	1.0000	0.5500	−1.2500	−0.5000	0.0000	0.5000	1,500.00
$z_j - c_j$		0.0000	0.0000	2.1375	3.2901	1.5900	0.0000	1.9030	12,697

An interesting aspect to this solution should be noted. The problem allowed for the choice of four different products to be manufactured. One might think that with three constraints, three products would be chosen for manufacture, but such is not the case as shown by the example solution. Only two products were chosen for manufacture. The third basic variable is the slack in machine center *B*.

It is clear that the pivot element was necessarily 1.4411, since it was necessary that the vector $\mathbf{y}_4$ be transformed into a unit vector with its second component equal to 1. Fortunately, the pivot element was a positive number. Had the element been a negative number, we would still have been required to use it for this method. The result of using a negative pivot element is that the corresponding solution will be nonfeasible. The process for eliminating the nonfeasibility is the dual-simplex method, which has been mentioned before but has not yet been discussed.

The preceding illustrations of the method of changing one of the original vectors in the **A** matrix were not particularly complex, but they were handled one vector at a time. The process would be the same if it were required to see the effect of changing several vectors. The problem becomes more complicated and it is doubtful whether there is any advantage in proceeding from the optimal solution rather than starting to solve the problem all over again from the beginning.

5-16 ADDITION OF A NEW CONSTRAINT

In considering the addition of a new constraint, the first thing to be done is to test whether the constraint is an active one or not. The test is made by direct substitution of the optimal solution into the constraint. If the constraint is violated, the procedure is illustrated by the following example. Consider our optimal solution to the example on p. 69. Suppose that we apply the following constraint:

$$x_1 + x_2 \leqq 1200$$

This is an active constraint since $x_1 + x_2 = 1794.11$. It is further evident that the maximum profit must be less than the 12,737 previously obtained. In order to incorporate the new constraint, we merely add the constraint to the optimal tableau after it has been made into the form of an equation by adding a slack variable x_8. We have enlarged the tableau (see page 124, top) to four constraints, so our identity matrix will be of the fourth order.

This tableau is a perfectly legitimate statement of the constraints but is not in a form in which the solutions can be read. We need to have four unit vectors of the fourth order. The current unit vectors x_1 and x_2 were destroyed since the variables x_1 and x_2 both appeared in the added constraint. We shall proceed to make the transformations that will convert the

c_j		5.24	7.30	8.34	4.18	0.00	0.00	0.00	0.00	000
Variables		x_1	x_2	x_3	x_4	x_5	x_6	x_7	x_8	b
$\mathbf{x_B}$	$\mathbf{c''_B}$									
x_1	5.24	1.0000	0.0000	1.9353	0.0000	0.6471	−0.2353	0.1765	0.0000	294.11
x_4	4.18	0.0000	0.0000	−1.0529	1.0000	0.5294	0.3529	−0.7647	0.0000	58.82
x_2	7.30	0.0000	1.0000	0.5500	0.0000	−0.5000	0.0000	0.5000	0.0000	1500.00
x_8	0.00	1.0000	1.0000	0.0000	0.0000	0.0000	0.0000	0.0000	1.0000	1200.00
$z_j - c_j$		—	—	—	—	—	—	—	—	—

vectors associated with x_1, x_2, x_4, and x_8 into unit vectors if they do not already appear as such. In that form we will then have an identity matrix, which is so necessary for the simplex procedure. To accomplish this it will be necessary to subtract the first row from the fourth row and then the third row from the fourth row. The $z_j - c_j$ is then computed and the result is:

c_j		5.24	7.30	8.34	4.18	0.00	0.00	0.00	0.00	000
Variables		x_1	x_2	x_3	x_4	x_5	x_6	x_7	x_8	b
$\mathbf{x_B}$	$\mathbf{c'_B}$									
x_1	5.24	1.0000	0.0000	1.9353	0.0000	0.6471	−0.2353	0.1765	0.0000	294.11
x_4	4.18	0.0000	0.0000	−1.0529	1.0000	0.5294	0.3529	−0.7647	0.0000	58.82
x_2	7.30	0.0000	1.0000	0.5500	0.0000	−0.5000	0.0000	0.5000	0.0000	1,500.00
x_8	0.00	0.0000	0.0000	−2.4853	0.0000	−0.1471	0.2353	−0.6765	1.0000	−594.11
$z_j - c_j$		0.0000	0.0000	1.4146	0.0000	1.9537	0.2423	1.3782	0.0000	12,737

Although we have converted this to a basic solution, it is no longer feasible. The value of x_8 is negative but the solution tests optimal since the entire $z_j - c_j$ row is nonnegative. At this point we are faced with a problem which has been mentioned before. We must remove the nonfeasibility without destroying the optimality. This is done by the *dual-simplex method*. It is rather obvious that a different basic solution must be found. To do this we first select the outgoing variable and from that determine the incoming variable, a process that is just the opposite of that used in the simplex procedure. The method has been called the dual-simplex algorithm since the simplex criteria have been applied to the dual rather than to the primal.

To remove the nonfeasibility, the entire fourth row must be divided by a *negative number*. If we scan the fourth row, we see that the elements in x_3, x_5, and x_7 are negative, and therefore if anyone of them is chosen as the pivot element, the nonfeasibility is removed. The selection must be made so that the resulting solution is still optimal (i.e., no term in the $z_j - c_j$ row can be allowed to turn negative). The selection is made by dividing each negative element in the row into the corresponding element in the $z_j - c_j$ row and selecting the one that gives a quotient with the smallest absolute value. In this case we find that -2.4853 will be the pivot point. The result of carrying out this iteration is:

c_j		5.24	7.30	8.34	4.18	0.00	0.00	0.00	0.00	000
Variables		x_1	x_2	x_3	x_4	x_5	x_6	x_7	x_8	b
$\mathbf{x_B}$	$\mathbf{c'_B}$									
x_1	5.24	1.0000	0.0000	0.0000	0.0000	0.5325	−0.0520	−0.3503	0.7787	−168.52
x_4	4.18	0.0000	0.0000	0.0000	1.0000	0.5917	0.2532	−0.4781	−0.4237	310.52
x_2	7.30	0.0000	1.0000	0.0000	0.0000	−0.5326	0.0521	0.3503	0.2213	1,368.52
x_3	8.34	0.0000	0.0000	1.0000	0.0000	0.0592	−0.0947	0.2722	−0.4024	239.05
$z_j - c_j$		0.0000	0.0000	0.0000	0.0000	1.8698	0.3763	0.9932	0.5693	12,399

The solution is still optimal but another nonfeasibility has appeared. The dual-simplex method is continued. x_1 will be removed and x_7 will be entered into the solution.

c_j		5.24	7.30	8.34	4.18	0.00	0.00	0.00	0.00	000
Variables		x_1	x_2	x_3	x_4	x_5	x_6	x_7	x_8	b
$\mathbf{x_B}$	$\mathbf{c'_B}$									
x_7	0.00	−2.8549	0.0000	0.0000	0.0000	−1.5203	0.1486	1.0000	−2.2230	481.07
x_4	4.18	−1.3649	0.0000	0.0000	1.0000	−0.1351	0.3243	0.0000	−1.4865	540.52
x_2	7.30	1.0000	1.0000	0.0000	0.0000	0.0000	0.0000	0.0000	1.0000	1,200.00
x_3	8.34	0.7771	0.0000	1.0000	0.0000	0.4730	−0.1351	0.0000	0.2027	108.10
$z_j - c_j$		2.8357	0.0000	0.0000	0.0000	3.3801	0.2284	0.0000	2.7769	11,920

Now the solution is both optimal and feasible. It can be seen that the optimal value has been reduced because of the added constraint.

5-17 REMOVAL OF A CONSTRAINT

The removal of a constraint from an optimal solution sometimes causes more trouble than the addition of a constraint. If the constraint is nonbinding, it can be removed and there is no effect on the solution. If the constraint is active, however, and this is seen by the fact that the dual variable corresponding to the constraint is nonzero, there is considerably more to the problem. Consider again the example stated at the beginning of the chapter. Suppose that we would like to remove the original third constraint.

Examination of the summary tableaux on p. 70, reveals that removal of the third constraint from the optimal tableau would also remove the variable x_2, since only zero components remain in that vector after the third constraint is wiped out. This would mean that x_2 could never again be a candidate for entry into the solution. This must be avoided.

What one would like to do is to remove the variable x_7 when the third constraint is removed. To do this the variable x_7 would have to be a basic variable at the time the constraint is removed. Therefore, starting from the optimal tableau, the vector x_7 is entered by the usual simplex procedure, with the exception that x_2 must be removed to have the 1 of the unit vector in the row of the constraint being removed. The resulting tableau is:

c_j		5.24	7.30	8.34	4.18	0.00	0.00	0.00	000
Variables		x_1	x_2	x_3	x_4	x_5	x_6	x_7	b
$\mathbf{x_B}$	$\mathbf{c'_B}$								
x_1	5.24	1.0000	−0.3530	1.7412	0.0000	0.8236	−0.2353	0.0000	−235.39
x_4	4.18	0.0000	1.5294	−0.2117	1.0000	−0.2353	0.3529	0.0000	2352.92
x_7	0.00	0.0000	2.0000	1.1000	0.0000	−1.0000	0.0000	1.0000	3000.00
$z_j - c_j$		0.0000	−2.7564	−0.1012	0.0000	3.3318	0.2423	0.0000	8602

In this tableau, with the third constraint and the seventh variable removed, we have a basic but nonfeasible solution:

c_j		5.24	7.30	8.34	4.18	0.00	0.00	0000
Variables		x_1	x_2	x_3	x_4	x_5	x_6	b
$\mathbf{x_B}$	$\mathbf{c'_B}$							
x_1	5.24	1.0000	−0.3530	1.7412	0.0000	0.8236	−0.2353	−235.39
x_4	4.18	0.0000	1.5294	−0.2117	1.0000	−0.2353	0.3529	2352.92
$z_j - c_j$		0.0000	−2.7564	−0.1012	0.0000	3.3318	0.2423	8602

This solution is also nonoptimal. Since it is neither nonoptimal nor nonfeasible, it does not lend itself very well to completion, even though it can be done. The nonfeasibility can be removed by using either the -0.3530 or the -0.2353 as a pivot and then using the regular simplex method to obtain optimality.

It is quite evident that in this case the removal of the constraint from the original problem and solving it completely from the beginning as a problem in two constraints and six variables is considerably more efficient than the process just described, although each is correct.

5-18 SUMMARY ON SENSITIVITY ANALYSIS

The discussion in this chapter on the analysis of a number of postoptimality problems is by no means complete nor is it intended to be. It does, however, present some of the more obvious analyses that one might wish to make. Such analyses after a problem has been solved may be needed either for direct decision making or for planning or even because some type of error has been made in original data and a correct result is needed quickly.

Another purpose or use of sensitivity analysis is associated with the original model building. Many people spend untold hours gathering the data for the original statement of the problem only to find that the solution to the problem was relatively insensitive to many of the data items. I suggest that, when formulating the problem, only reasonable care be used—to model a problem that is reasonably correct, solve the problem, and use sensitivity analysis to determine the items of data that need further attention. Many hours can be saved this way.

The material that has just been presented has been handled in a very straightforward manner and might lead us to believe that no problems ever arise in the solution of linear programming problems or in the sensitivity analyses of the same. The problems that do arise are usually not difficult to handle but would require considerably more theoretical development in order to give a full explanation and justification. Some of these problems arise in connection with degenerate solutions. It is felt that a more detailed exposition is not warranted in a textbook of this type.

It is significant, however, that reasonably sophisticated computer codes for solving linear programming problems recognize many of the problems mentioned above and give a message to the problem solver. Redundancy is no problem and generally need not be mentioned. Inconsistency is always recognized. For example, if the artificial variables cannot be eliminated, the problem is obviously inconsistent—when this occurs, the user is given such a message. Because of the sophistication of computer codes, we merely approach the solution of a problem as if the solution exists; if it does, we will obtain the solution, and if it does not, we are given a message.

EXERCISES

The exercises at the end of this chapter are designed to give practice in the formal aspects of sensitivity analysis. For the most part, reference will be made to solutions of exercises in previous chapters. The problems will necessarily be rather small since to this point we have not discussed computer solutions. We do not want to make the computations a burden.

1 Consider Exercise 1, Chap. 3 (p. 81). Its statement was:
Find

$$x_1, x_2 \geqq 0$$

such that

$$\begin{aligned} 2x_1 - x_2 &\leqq 8 \\ 2x_1 + x_2 &\leqq 12 \\ x_1 + 2x_2 &\leqq 16 \end{aligned}$$

and such that

$$z = 4x_1 + 3x_2 \text{ is a } maximum$$

(*a*) Find the dual solutions to this problem; that is, find the marginal value of an additional unit of resource for each of the right-hand sides. Which constraints are binding? If a constraint is binding, find the complete range for which the marginal rate holds; if it is not binding, find how many units of resource could be taken from the right-hand side before the constraint becomes binding.

(*b*) Examine the coefficients of the objective function and find the range over which each of these coefficients can move without changing the basic variables in the solution. Taken separately, what is the range over which the objective function will vary for the corresponding range of the coefficient?

2 Consider Exercise 2, Chap. 3 (p. 82). Its statement was:
Find

$$x_1, x_2, x_3 \geqq 0$$

such that

$$\begin{aligned} 3x_1 + 2x_2 - x_3 &\leqq 610 \\ -x_1 + 6x_2 + 3x_3 &\leqq 125 \\ -2x_1 + x_2 + 0.5x_3 &\leqq 420 \end{aligned}$$

and such that

$$z = 2x_1 + 5x_2 + 8x_3 \text{ is a } maximum$$

(*a*) Find the dual solutions; that is, find the marginal value of an additional unit of resource for each of the right-hand sides. Which constraints are binding? If a constraint is binding, find the complete range for which the marginal rate holds; if it is not binding, find how many units of resource could be taken away from the right-hand side before the constraint becomes binding.

(*b*) Examine the coefficients of the objective function and find the range over which each of these coefficients can move without changing the basic variables in the solution—in fact, without changing the solution values. Taken separately, what is the range over which the objective function will vary for the corresponding range of each of the coefficients?

3 Consider the product-mix problem solved in Exercise 3, Chap. 3 (p. 82).

(*a*) Find the amount that you would be willing to pay for an additional minute of time in each department: cutting, assembly, and testing. How many minutes would you be willing to purchase at that price?

(*b*) One of the three departments has idle time. If you could transfer this time to other departments, how much would you transfer to each of the other departments (do not go beyond upper limit), and what is the total increase in sales? How many boxes and how many crates would then be manufactured?

(*c*) By how much could the selling price on boxes drop before the optimal product mix would change? Answer the same for crates.

(*d*) A new container has been designed. It requires 2 min in cutting, 1 min in assembly, and 3 min in testing. How much would the selling price have to be in order for it to be in the optimal product mix?

4 Consider the gasoline blending problem given in Exercise 4, Chap. 3 (p. 82). There are six constraints to the problem. Which of these are binding and which are nonbinding?

(*a*) Give the marginal values of the resources in case of the binding constraints. For what range of values of the right-hand sides will these marginal values hold?

(*b*) By how much would the profit on regular gas from crude 1 have to increase before it would come into the solution?

(*c*) By how much would the profit on special gas from crude 1 have to drop before it would change the solution? Would special from crude 1 be removed?

5 Consider Exercise 10, Chap. 3 (p. 82), on TV and stereo production.

(*a*) Which of the four constraints are binding? Give the marginal values of the binding constraints. Give the upper limit and total gain in profit to be made by increasing each of the right-hand sides to that upper limit for those binding constraints.

(*b*) How is the solution affected if the profit on the stereo set is reduced to \$40?

(*c*) What is the effect on the solution if the constraint on speaker production is removed?

(*d*) How is the optimal solution affected if the requirement is added that the number of stereo sets produced must be no greater than one-sixth of the number of TV sets produced? Carry out the computation far enough to find out how many of each should be produced under this constraint. You need not carry out all the computations for the entire tableau if you can establish optimality.

6 Consider the sweeper and polisher problem originally presented in Exercise 4, Chap. 1 (p. 12), whose simplex solution was called for in Exercise 4, Chap. 4 (p. 100).

Referring to the simplex solution, examine the final tableau for the case in which the profit on the polisher is 50 cents.

(*a*) From the optimal solution, do a sensitivity analysis to see if the five extra runs should be rented.

(*b*) Move the profit on the polisher up to \$1.50 to see if the basis changes. If it goes out of range, perform an iteration and then do a sensitivity analysis on the coefficient of the variable for the polisher.

(*c*) Find the solution to the problem for the model that has a variable representing the number of extra runs and then do a sensitivity analysis on the coefficient of the variable for the polisher.

7 Referring to the simplex solution of Exercise 3, Chap. 4 (p. 100), do a complete sensitivity analysis on both the right-hand sides and on the coefficients of the objective function. Note that this is a minimization problem.

CHAPTER

SIX

REVISED SIMPLEX PROCEDURE†

6-1 INTRODUCTION

Earlier in the text, reference was made to the use of the computer for solving linear programming problems. In most cases, the simplex procedure is not coded directly for solving such problems. A variation of it known as the *revised simplex procedure* is most often used.

The "revised" aspect concerns the procedure for changing or updating the tableaux. From the past discussions, we noted that it was necessary to examine the $z_j - c_j$ row in order to test for optimality and if not optimal, the row was used to select the next incoming vector; thus we must be able to calculate $z_j - c_j$, which is done by computing $\mathbf{c_B}\,\mathbf{y}_j - c_j$, where the $\mathbf{y}_j$ refers to the updated $\mathbf{a}_j$ in the tableaux being examined. $\mathbf{y}_j$ itself can be computed by multiplying $\mathbf{B}^{-1}$ times the original $\mathbf{a}_j$. Thus to be able to compute the current $z_j - c_j$, we need to know $\mathbf{B}^{-1}$, the inverse of the basis set of vectors; $\mathbf{c_B}$, the coefficients of the objective function for the vectors currently in the basis; $\mathbf{a}_j$, the set of vectors in the initial tableau; and c_j, the coefficients of the entire objective function. In order to be aware of the current solution, and the current value of the objective function, one needs to compute

$$\mathbf{x_B} = \mathbf{B}^{-1}\mathbf{b} \qquad \text{and} \qquad z = \mathbf{c_B}\,\mathbf{x_B} = \mathbf{c_B}\,\mathbf{B}^{-1}\mathbf{b}$$

† The example used in this chapter is the same as that of Chap. 3. There may be slight variations in the last decimal position of some of the numbers. These differences are due to the number of decimals carried in the computations and the resulting rounding differences.

We really do not need to have the entire updated tableau, but only a sufficient part of it to be able to construct the items mentioned above. The process requires that we know the initial tableau and $\mathbf{B}^{-1}$, the inverse of the basis set of vectors at each stage. To have $\mathbf{B}^{-1}$ at each stage, we need the ability to calculate $\mathbf{B}^{-1}$ from a previous inverse when only one of the vectors is changed. The method of doing this was developed in Sec. 2-20, p. 45, in the discussion of the product form of the inverse.

6-2 REVISED TABLEAU

Let us now develop the revised simplex approach by referring again to the example we have been using. Recalling that the objective function can be considered as just another constraint, let us write it as the first constraint so that our initial tableau becomes as shown below.

Initial tableau

z	x_1	x_2	x_3	x_4	x_5	x_6	x_7	b
1.0000	−5.2400	−7.3000	−8.3400	−4.1800	0.0000	0.0000	0.0000	0000
0.0000	1.5000	1.0000	2.4000	1.0000	1.0000	0.0000	0.0000	2000.00
0.0000	1.0000	5.0000	1.0000	3.5000	0.0000	1.0000	0.0000	8000.00
0.0000	1.5000	3.0000	3.5000	1.0000	0.0000	0.0000	1.0000	5000.00

The basis set of vectors in this form are those corresponding to z, x_5, x_6, and x_7, and it will be denoted by $\mathbf{B}_1$. Its inverse will be denoted by $\mathbf{B}_1^{-1}$, and initially we have

$$\mathbf{B}_1^{-1} = \begin{bmatrix} 1 & 0 & 0 & 0 \\ 0 & 1 & 0 & 0 \\ 0 & 0 & 1 & 0 \\ 0 & 0 & 0 & 1 \end{bmatrix}$$

In general, we can write

$$\mathbf{B}_1 = \begin{bmatrix} 1 & -\mathbf{c_B} \\ 0 & \mathbf{B} \end{bmatrix} \quad \text{and therefore} \quad \mathbf{B}_1^{-1} = \begin{bmatrix} 1 & \mathbf{c_B}\mathbf{B}^{-1} \\ 0 & \mathbf{B}^{-1} \end{bmatrix}$$

where $\mathbf{B}$ and $\mathbf{B}^{-1}$ refer to the basis set of vectors in the regular simplex approach to the problem. Although I have not developed the inverse of a matrix in a partitioned form, it is easy to see that as defined above,

$$\mathbf{B}_1\mathbf{B}_1^{-1} = \begin{bmatrix} 1 & 0 \\ 0 & \mathbf{I}_m \end{bmatrix}$$

which is an identity matrix of order $m + 1$.

Again using the partitioned form of the matrices, our problem can be written:

Find $\mathbf{x} \geqq \mathbf{0}$ such that

$$\begin{bmatrix} 1 & -\mathbf{c} \\ 0 & \mathbf{A} \end{bmatrix} \begin{bmatrix} z \\ \mathbf{x} \end{bmatrix} = \begin{bmatrix} 0 \\ \mathbf{b} \end{bmatrix}$$

and z (unrestricted in sign) is a *maximum*

When expanded, this becomes

Find

$$\mathbf{x} \geqq \mathbf{0}$$

such that

$$z - \mathbf{cx} = 0$$

$$\mathbf{Ax} = \mathbf{b}$$

and such that z is a *maximum*

6-3 SELECTION OF INCOMING VARIABLE

Since each column of the matrix of coefficients when written in this form can be written

$$\begin{bmatrix} -c_j \\ \mathbf{a}_j \end{bmatrix}$$

we can update any column by multiplying that column from the initial tableau by $\mathbf{B}_1^{-1}$.

Thus,

$$\mathbf{B}_1^{-1} \begin{bmatrix} -\mathbf{c}_j \\ \mathbf{a}_j \end{bmatrix} = \begin{bmatrix} 1 & \mathbf{c_B B}^{-1} \\ 0 & \mathbf{B}^{-1} \end{bmatrix} \begin{bmatrix} -c_j \\ \mathbf{a}_j \end{bmatrix} = \begin{bmatrix} -c_j + \mathbf{c_B B}^{-1}\mathbf{a}_j \\ \mathbf{B}^{-1}\mathbf{a}_j \end{bmatrix} = \begin{bmatrix} z_j - c_j \\ \mathbf{y}_j \end{bmatrix}$$

In this final form the first row is the familiar $z_j - c_j$, and the $\mathbf{y}_j$ represents the rest of the column (i.e., it is a vector with as many components as there are constraints in the original problem). Applying this to our example problem, we first wish to find the incoming vector by finding the first row in the product above, and from it selecting the largest negative. The first row of $\mathbf{B}_1^{-1}$ in this example is (1, 0, 0, 0) and in multiplying this by

$$\begin{bmatrix} -c_j \\ \mathbf{a}_j \end{bmatrix}$$

we obtain the vector

$$(1,\ -5.24,\ -7.30,\ \overset{*}{-8.34},\ -4.18,\ 0,\ 0,\ 0,\ 0) \qquad \text{(6-1)}$$

The first component will always be 1 since it corresponds to z, which will always be kept in the basis. The last element of (6-1) is initially 0 and corresponds to the current value of the objective function, since we have in general, for the last element,

$$(1, \mathbf{c_B}\mathbf{B}^{-1})\begin{bmatrix}0\\ \mathbf{b}\end{bmatrix} = \mathbf{c_B}\mathbf{B}^{-1}\mathbf{b} = \mathbf{c_B}\mathbf{x_B} = z$$

The other three zeros in (6-1) correspond to the basic variables x_5, x_6, and x_7. In order to adjust our notation to this form of the problem, we shall let

$$\mathbf{a}_j^{(\mathrm{R})} = \begin{bmatrix}-c_j\\ \mathbf{a}_j\end{bmatrix} \quad \text{and} \quad \mathbf{y}_j^{(\mathrm{R})} = \begin{bmatrix}z_j - c_j\\ \mathbf{y}_j\end{bmatrix}$$

and this $\mathbf{y}_j^{(\mathrm{R})} = \mathbf{B}_1^{-1}\mathbf{a}_j^{(\mathrm{R})}$. The superscript (R) indicates that we are using the revised simplex form of the tableaux. The selection of the incoming vector is made by examining the first row of $\mathbf{B}_1^{-1}\mathbf{a}_j^{(\mathrm{R})}$, after eliminating the first and last components. The incoming vector corresponds to the minimum component, which in this example is -8.34,† and it corresponds to the variable x_3. We now need to insert x_3 in place of x_5, x_6, or x_7. The variable z is not a candidate for being removed. We need to know the remainder of the components of the vector $\mathbf{y}_3^{(\mathrm{R})}$, having already obtained the first component, -8.34. We also need to know the components of $\mathbf{x_B}^{(\mathrm{R})}$ and the value of the objective function, which are given by $\mathbf{B}_1^{-1}\mathbf{b}^{(\mathrm{R})}$. Thus we have

$$\mathbf{y}_3^{(\mathrm{R})} = \mathbf{B}_1^{-1}\mathbf{a}_3^{(\mathrm{R})} = \begin{bmatrix}1&0&0&0\\0&1&0&0\\0&0&1&0\\0&0&0&1\end{bmatrix}\begin{bmatrix}\overset{*}{-8.34}\\2.40\\1.00\\3.50\end{bmatrix} = \begin{bmatrix}-8.34\\2.40\\1.00\\3.50\end{bmatrix}$$

and

$$\mathbf{x_B}^{(\mathrm{R})} = \mathbf{B}_1^{-1}\mathbf{b}^{(\mathrm{R})} = \begin{bmatrix}1&0&0&0\\0&1&0&0\\0&0&1&0\\0&0&0&1\end{bmatrix}\begin{bmatrix}0000\\2000\\8000\\5000\end{bmatrix} = \begin{bmatrix}0000\\2000\\8000\\5000\end{bmatrix}$$

6-4 SELECTION OF OUTGOING VARIABLE

In order to determine the outgoing variable, we must divide the components of $\mathbf{x_B}^{(\mathrm{R})}$ by the corresponding components of $\mathbf{y}_3^{(\mathrm{K})}$ and select the smallest of these ratios, just as we did in the regular simplex method. Since $2000/2.4 = 833.33$ is smaller than either of $8000/1$ or $5000/3.5$, and since the 2000 corre-

† Below we use an asterisk to identify the entering variable.

sponds to the variable x_5, it is the one to be removed. This means that the new set of basis variables will be z, x_3, x_6, and x_7. From our work in Chap. 2, p. 47, the new basis inverse will be $\mathbf{EB}_1^{-1}$, where

$$\mathbf{E} = \begin{bmatrix} 1 & 8.34/2.4 & 0 & 0 \\ 0 & 1/2.4 & 0 & 0 \\ 0 & -1/2.4 & 1 & 0 \\ 0 & -3.5/2.4 & 0 & 1 \end{bmatrix} = \begin{bmatrix} 1 & 3.4750 & 0 & 0 \\ 0 & 0.4167 & 0 & 0 \\ 0 & -0.4167 & 1 & 0 \\ 0 & -1.4583 & 0 & 1 \end{bmatrix}$$

6-5 UPDATED BASIS INVERSE

Since we are starting with $\mathbf{B}_1$ as an identity matrix, the new inverse, which we shall denote by $\hat{\mathbf{B}}_1^{-1}$, will be just E as shown above. The algorithm for finding $\mathbf{E}$ as developed in Chap. 2 is as follows.

Since the vector

$$\mathbf{a}_3^{(R)} = \begin{bmatrix} -8.34 \\ 2.40 \\ 1.00 \\ 3.50 \end{bmatrix}$$

is to replace the *second* basis vector, the matrix E is obtained by starting with an identity matrix and altering its *second* column. The *second* element of $\mathbf{a}_3^{(R)}$ is replaced with its reciprocal, 1/2.4. Each of the other components is divided by that *second* element, 2.4, and then reversed in sign. Thus we have

$$\hat{\mathbf{B}}_1^{-1} = \mathbf{EB}_1^{-1} = \mathbf{EI} = \begin{bmatrix} 1 & 3.4750 & 0 & 0 \\ 0 & 0.4167 & 0 & 0 \\ 0 & -0.4167 & 1 & 0 \\ 0 & -1.4583 & 0 & 1 \end{bmatrix}$$

and the new $\hat{\mathbf{x}}_{\mathbf{B}}^{(R)}$ is

$$\mathbf{B}_1^{-1}\begin{bmatrix} 0 \\ \mathbf{b} \end{bmatrix} = \mathbf{B}_1^{-1}\mathbf{b}^{(R)} = \hat{\mathbf{x}}_{\mathbf{B}}^{(R)} = \begin{bmatrix} 1 & 3.4750 & 0 & 0 \\ 0 & 0.4167 & 0 & 0 \\ 0 & -0.4167 & 1 & 0 \\ 0 & -1.4583 & 0 & 1 \end{bmatrix} \begin{bmatrix} 0000 \\ 2000 \\ 8000 \\ 5000 \end{bmatrix} = \begin{bmatrix} 6950 \\ 833.40 \\ 7166.60 \\ 2083.40 \end{bmatrix}$$

which shows that $z = \$6950$, $x_3 = 833.40$, $x_6 = 7166.60$, and $x_7 = 2083.40$. Since we now have the new basis inverse, we can repeat the operation and continue to optimality.

6-6 CONDENSED TABLEAU

The format above is a little inconvenient, so let us designate the columns of the inverse, in order, as $\boldsymbol{\beta}_0$, $\boldsymbol{\beta}_1$, $\boldsymbol{\beta}_2$, and $\boldsymbol{\beta}_3$ for this fourth-order case. $\boldsymbol{\beta}_0$ is always a unit vector with the 1 in the first position, so it really does not have to be recorded, but we shall do so for completeness. Our initial brief summary tableau will then be:

Condensed initial tableau

Variables in basis	$\boldsymbol{\beta}_0$	$\boldsymbol{\beta}_1$	$\boldsymbol{\beta}_2$	$\boldsymbol{\beta}_3$	$\mathbf{x}_\mathbf{B}^{(R)}$	$\mathbf{y}_3^{(R)}$
z	1	0	0	0	0000	−8.34
x_5	0	1	0	0	(2000)	(2.40)
x_6	0	0	1	0	8000	1.00
x_7	0	0	0	1	5000	3.50

All of the information above has been recorded from the initial tableau shown on p. 131, with the exception of the $\mathbf{y}_3^{(R)}$ column, which has been computed and then recorded. As indicated before, the vector $\mathbf{y}_k^{(R)}$, in this case $k = 3$, has been selected after computing the scalar product of the first row of $\mathbf{B}_1^{-1}$ with the $\mathbf{a}_j^{(R)}$'s not in the basis and selecting the minimum value. In this case the minimum value of -8.34 corresponds to the variable x_3, so then $\mathbf{B}_1^{-1}$ was multiplied by the entire vector $\mathbf{a}_3^{(R)}$ to determine the entire vector $\mathbf{y}_3^{(R)}$, as shown in the right-hand column of the summary tableau above. The positive terms of $\mathbf{y}_3^{(R)}$ are divided term by term into the components of $\mathbf{x}_\mathbf{B}^{(R)}$ to find the minimum value, which corresponds to the basic variable x_5, so it will be the outgoing variable.

6-7 CONDENSED SECOND TABLEAU

The updated inverse was computed on p. 134, and it is shown below with the proper labels.

Second tableau

Variables in basis	$\boldsymbol{\beta}_0$	$\boldsymbol{\beta}_1$	$\boldsymbol{\beta}_2$	$\boldsymbol{\beta}_3$	$\mathbf{x}_\mathbf{B}^{(R)}$	$\mathbf{y}_k^{(R)}$
z	1	3.4750	0	0		
x_3	0	0.4167	0	0		
x_6	0	−0.4167	1	0		
x_7	0	−1.4583	0	1		

The last two columns, which have been left blank, are computed as follows:

$$\hat{\mathbf{x}}_{\mathbf{B}}^{(R)} = \hat{\mathbf{B}}_1^{-1}\mathbf{b}^{(R)} = \begin{bmatrix} 1 & 3.4750 & 0 & 0 \\ 0 & 0.4167 & 0 & 0 \\ 0 & -0.4167 & 1 & 0 \\ 0 & -1.4583 & 0 & 1 \end{bmatrix} \begin{bmatrix} 0000 \\ 2000 \\ 8000 \\ 5000 \end{bmatrix} = \begin{bmatrix} 6950 \\ 833.40 \\ 7166.60 \\ 2083.40 \end{bmatrix}$$

In order to determine $\mathbf{y}_k^{(R)}$, we first determine k by computing $z_j - c_j$ for all the nonbasic variables. This is done by selecting the first row of $\hat{\mathbf{B}}_1^{-1}$ shown above and multiplying by the vectors for the nonbasic variables. The result is

$$(1, 3.4750, 0, 0)(\mathbf{a}_1^{(R)}, \mathbf{a}_2^{(R)}, \mathbf{a}_4^{(R)}, \mathbf{a}_5^{(R)})$$
$$= (-0.0275, -3.8250, -0.7050, 3.4750)$$

The term -3.8250, the negative with the largest absolute value, corresponds to $\mathbf{a}_2^{(R)}$ (i.e., $k = 2$) and we now compute $\mathbf{y}_2^{(R)}$ for entry into the table above.

$$\mathbf{y}_2^{(R)} = \hat{\mathbf{B}}_1^{-1}\mathbf{a}_2^{(R)} = \begin{bmatrix} 1 & 3.4750 & 0 & 0 \\ 0 & 0.4167 & 0 & 0 \\ 0 & -0.4167 & 1 & 0 \\ 0 & -1.4583 & 0 & 1 \end{bmatrix} \begin{bmatrix} -7.30 \\ 1.00 \\ 5.00 \\ 3.00 \end{bmatrix} = \begin{bmatrix} -3.8250 \\ 0.4167 \\ 4.5833 \\ 1.5417 \end{bmatrix}$$

and the new full condensed tableau is

Condensed second tableau

Variables in basis	β_0	β_1	β_2	β_3	$\mathbf{x}_{\mathbf{B}}^{(R)}$	$\mathbf{y}_2^{(R)}$
z	1	3.4750	0	0	6950	−3.8250
x_3	0	0.4167	0	0	833.40	0.4167
x_6	0	−0.4167	1	0	7166.60	4.5833
x_7	0	−1.4583	0	1	(2083.40)	(1.5417)

In the process of updating the inverse, we used the notation $\hat{\mathbf{B}}_1^{-1}$ and $\hat{\mathbf{x}}_{\mathbf{B}}^{(R)}$ in order to distinguish them from $\mathbf{B}_1^{-1}$ and $\mathbf{x}_{\mathbf{B}}$. Once the computations have been made, we revert back to $\mathbf{B}_1^{-1}$ in preparation for another updating.

Division of $\mathbf{x}_{\mathbf{B}}^{(R)}$ by $\mathbf{y}_2^{(R)}$ term by term shows that the variable x_7 is to be removed, since the smallest quotient is 2083.40/1.5417, which yields 1351.37. We now need a new inverse with x_2 replacing x_7 in the basis. For this,

$$\mathbf{E} = \begin{bmatrix} 1 & 0 & 0 & 3.8250/1.5417 \\ 0 & 1 & 0 & -0.4167/1.5417 \\ 0 & 0 & 1 & -4.5833/1.5417 \\ 0 & 0 & 0 & 1/1.5417 \end{bmatrix} = \begin{bmatrix} 1 & 0 & 0 & 2.4810 \\ 0 & 1 & 0 & -0.2703 \\ 0 & 0 & 1 & -2.9729 \\ 0 & 0 & 0 & 0.6486 \end{bmatrix}$$

From this we then have

$$\hat{\mathbf{B}}_1^{-1} = \mathbf{E}\mathbf{B}_1^{-1} = \begin{bmatrix} 1 & 0 & 0 & 2.4810 \\ 0 & 1 & 0 & -0.2703 \\ 0 & 0 & 1 & -2.9729 \\ 0 & 0 & 0 & 0.6486 \end{bmatrix} \begin{bmatrix} 1 & 3.4750 & 0 & 0 \\ 0 & 0.4167 & 0 & 0 \\ 0 & -0.4167 & 1 & 0 \\ 0 & -1.4583 & 0 & 1 \end{bmatrix}$$

$$= \begin{bmatrix} 1 & -0.1430 & 0 & 2.4810 \\ 0 & 0.8109 & 0 & -0.2703 \\ 0 & 3.9187 & 1 & -2.9729 \\ 0 & -0.9459 & 0 & 0.6486 \end{bmatrix}$$

6-8 CONDENSED THIRD TABLEAU

We now have

Third tableau

Variables in basis	$\boldsymbol{\beta}_0$	$\boldsymbol{\beta}_1$	$\boldsymbol{\beta}_2$	$\boldsymbol{\beta}_3$	$\mathbf{x}_\mathbf{B}^{(R)}$	$\mathbf{y}_k^{(R)}$
z	1	−0.1430	0	2.4810		
x_3	0	0.8109	0	−0.2703		
x_6	0	3.9187	1	−2.9729		
x_2	0	−0.9459	0	0.6486		

From this we find

$$\hat{\mathbf{x}}_\mathbf{B}^{(R)} = \hat{\mathbf{B}}_1^{-1}\mathbf{b}^{(R)} = \begin{bmatrix} 1 & -0.1430 & 0 & 2.4810 \\ 0 & 0.8109 & 0 & -0.2703 \\ 0 & 3.9187 & 1 & -2.9729 \\ 0 & -0.9459 & 0 & 0.6486 \end{bmatrix} \begin{bmatrix} 0000 \\ 2000 \\ 8000 \\ 5000 \end{bmatrix} = \begin{bmatrix} 12119 \\ 270.30 \\ 973.00 \\ 1351.20 \end{bmatrix}$$

Computing $z_j - c_j$ for all of the nonbasic variables, we select the first row of the current $\hat{\mathbf{B}}_1^{-1}$, and by multiplying by the vectors for the nonbasic variables, we have

$$(1, -0.1430, 0, 2.4810)(\mathbf{a}_1^{(R)}, \mathbf{a}_4^{(R)}, \mathbf{a}_5^{(R)}, \mathbf{a}_7^{(R)})$$
$$= (-1.7330, -1.8420, -0.1430, 2.4810)$$

The term -1.8420, the negative with the largest absolute value, corresponds to $\mathbf{a}_4^{(R)}$ (i.e., $k = 4$), and therefore

$$\mathbf{y}_4^{(R)} = \hat{\mathbf{B}}_1^{-1}\mathbf{a}_4^{(R)} = \begin{bmatrix} 1 & -0.1430 & 0 & 2.4810 \\ 0 & 0.8109 & 0 & -0.2703 \\ 0 & 3.9187 & 1 & -2.9729 \\ 0 & -0.9459 & 0 & 0.6486 \end{bmatrix} \begin{bmatrix} -4.18 \\ 1.00 \\ 3.50 \\ 1.00 \end{bmatrix} = \begin{bmatrix} -1.8420 \\ 0.5406 \\ 4.4458 \\ -0.2973 \end{bmatrix}$$

and the complete condensed third tableau becomes:

Condensed third tableau

Variables in basis	β_0	β_1	β_2	β_3	$\mathbf{x}_\mathbf{B}^{(R)}$	$\mathbf{y}_4^{(R)}$
z	1	−0.1430	0	2.4810	12119	−1.8420
x_3	0	0.8109	0	−0.2703	270.30	0.5406
x_6	0	3.9187	1	−2.9729	(973.00)	(4.4458)
x_2	0	−0.9459	0	0.6486	1351.20	−0.2973

Division of $\mathbf{x}_\mathbf{B}^{(R)}$ by $\mathbf{y}_4^{(R)}$ term by term shows that variable x_6 is to be removed since the smallest quotient is 973/4.4458, which yields 218.85. We now need a new inverse with x_4 replacing x_6 in the basis. For this

$$\mathbf{E} = \begin{bmatrix} 1 & 0 & 1.8420/4.4458 & 0 \\ 0 & 1 & -0.5406/4.4458 & 0 \\ 0 & 0 & 1/4.4458 & 0 \\ 0 & 0 & 0.2973/4.4458 & 1 \end{bmatrix} = \begin{bmatrix} 1 & 0 & 0.4144 & 0 \\ 0 & 1 & -0.1216 & 0 \\ 0 & 0 & 0.2249 & 0 \\ 0 & 0 & 0.0669 & 1 \end{bmatrix}$$

and

$$\hat{\mathbf{B}}_1^{-1} = \mathbf{E}\mathbf{B}_1^{-1} = \begin{bmatrix} 1 & 0 & 0.4144 & 0 \\ 0 & 1 & -0.1216 & 0 \\ 0 & 0 & 0.2249 & 0 \\ 0 & 0 & 0.0669 & 1 \end{bmatrix} \begin{bmatrix} 1 & -0.1430 & 0 & 2.4810 \\ 0 & 0.8109 & 0 & -0.2703 \\ 0 & 3.9187 & 1 & -2.9729 \\ 0 & -0.9459 & 0 & 0.6486 \end{bmatrix}$$

$$= \begin{bmatrix} 1 & 1.4809 & 0.4144 & 1.2490 \\ 0 & 0.3344 & -0.1216 & 0.0912 \\ 0 & 0.8813 & 0.2249 & -0.6686 \\ 0 & -0.6837 & 0.0669 & 0.4497 \end{bmatrix}$$

6-9 CONDENSED FOURTH TABLEAU

We now have the

Fourth tableau

Variables in basis	β_0	β_1	β_2	β_3	$\mathbf{x}_\mathbf{B}^{(R)}$	$\mathbf{y}_k^{(R)}$
z	1	1.4809	0.4144	1.2490		
x_3	0	0.3344	−0.1216	0.0912		
x_4	0	0.8813	0.2249	−0.6686		
x_2	0	−0.6837	0.0669	0.4497		

From this we find

$$\hat{\mathbf{x}}_{\mathbf{B}} = \hat{\mathbf{B}}_1^{-1}\mathbf{b}^{(R)} = \begin{bmatrix} 1 & 1.4809 & 0.4144 & 1.2490 \\ 0 & 0.3344 & -0.1216 & 0.0912 \\ 0 & 0.8813 & 0.2249 & -0.6686 \\ 0 & -0.6837 & 0.0669 & 0.4497 \end{bmatrix} \begin{bmatrix} 0000 \\ 2000 \\ 8000 \\ 5000 \end{bmatrix} = \begin{bmatrix} 12522 \\ 152.00 \\ 218.80 \\ 1416.30 \end{bmatrix}$$

Computing $z_j - c_j$ for all of the nonbasic variables, we select the first row of the current $\hat{\mathbf{B}}_1^{-1}$, and by multiplying by the vectors for the nonbasic vectors, we have

$$(1, 1.4809, 0.4144, 1.2490)(\mathbf{a}_1^{(R)}, \mathbf{a}_5^{(R)}, \mathbf{a}_6^{(R)}, \mathbf{a}_7^{(R)})$$
$$= (-0.7308, 1.4809, 0.4144, 1.2490)$$

The term -0.7308, the negative with the largest absolute value, corresponds to $\mathbf{a}_1^{(R)}$ (i.e., $k = 1$) and therefore

$$\mathbf{y}_1^{(R)} = \hat{\mathbf{B}}_1^{-1}\mathbf{a}_1^{(R)} = \begin{bmatrix} 1 & 1.4809 & 0.4144 & 1.2490 \\ 0 & 0.3344 & -0.1216 & 0.0912 \\ 0 & 0.8813 & 0.2249 & -0.6686 \\ 0 & -0.6837 & 0.0669 & 0.4497 \end{bmatrix} \begin{bmatrix} -5.24 \\ 1.50 \\ 1.00 \\ 1.50 \end{bmatrix}$$

$$= \begin{bmatrix} -0.7308 \\ 0.5168 \\ 0.5440 \\ -0.2841 \end{bmatrix}$$

and the complete condensed fourth tableau is:

Condensed fourth tableau

Variables in basis	β_0	β_1	β_2	β_3	$\mathbf{x}_{\mathbf{B}}^{(R)}$	$\mathbf{y}_1^{(R)}$
z	1	1.4809	0.4144	1.2490	12522	−0.7308
x_3	0	0.3344	−0.1216	0.0912	(152.00)	(0.5168)
x_4	0	0.8813	0.2249	−0.6686	218.80	0.5440
x_2	0	−0.6837	0.0669	0.4497	1416.30	−0.2841

Division of $\mathbf{x}_{\mathbf{B}}^{(R)}$ by $\mathbf{y}_1^{(R)}$ term by term shows that the variable x_3 is to be removed since the smallest quotient is 152/0.5168, which equals 294.12. We now need a new inverse, with x_1 replacing x_3 in the basis.

$$\mathbf{E} = \begin{bmatrix} 1 & 0.7308/0.5168 & 0 & 0 \\ 0 & 1/0.5168 & 0 & 0 \\ 0 & -0.5440/0.5168 & 1 & 0 \\ 0 & 0.2841/0.5168 & 0 & 1 \end{bmatrix} = \begin{bmatrix} 1 & 1.4143 & 0 & 0 \\ 0 & 1.9350 & 0 & 0 \\ 0 & -1.0526 & 1 & 0 \\ 0 & 0.5499 & 0 & 1 \end{bmatrix}$$

and

$$\hat{\mathbf{B}}_1^{-1} = \mathbf{E}\mathbf{B}_1^{-1}$$

$$= \begin{bmatrix} 1 & 1.4143 & 0 & 0 \\ 0 & 1.9350 & 0 & 0 \\ 0 & -1.0526 & 1 & 0 \\ 0 & 0.5499 & 0 & 1 \end{bmatrix} \begin{bmatrix} 1 & 1.4809 & 0.4144 & 1.2490 \\ 0 & 0.3344 & -0.1216 & 0.0912 \\ 0 & 0.8813 & 0.2249 & -0.6686 \\ 0 & -0.6837 & 0.0669 & 0.4497 \end{bmatrix}$$

$$= \begin{bmatrix} 1 & 1.9538 & 0.2423 & 1.3780 \\ 0 & 0.6471 & -0.2353 & 0.1765 \\ 0 & 0.5293 & 0.3529 & -0.7646 \\ 0 & -0.5000 & 0.0000 & 0.5000 \end{bmatrix}$$

6-10 CONDENSED OPTIMAL TABLEAU

We now have the

Condensed optimal tableau

Variables in basis	β_0	β_1	β_2	β_3	$\mathbf{x}_\mathbf{B}^{(R)}$	$\mathbf{y}_k^{(R)}$
z	1	1.9538	0.2423	1.3780		
x_1	0	0.6471	−0.2353	0.1765		
x_4	0	0.5293	0.3529	−0.7646		
x_2	0	−0.5000	0.0000	0.5000		

From this we find

$$\hat{\mathbf{x}}_\mathbf{B} = \hat{\mathbf{B}}_1^{-1}\mathbf{b}^{(R)} = \begin{bmatrix} 1 & 1.9538 & 0.2423 & 1.3780 \\ 0 & 0.6471 & -0.2353 & 0.1765 \\ 0 & 0.5293 & 0.3529 & -0.7646 \\ 0 & -0.5000 & 0.0000 & 0.5000 \end{bmatrix} \begin{bmatrix} 0000 \\ 2000 \\ 8000 \\ 5000 \end{bmatrix}$$

$$= \begin{bmatrix} 12737 \\ 294.30 \\ 58.80 \\ 1500.00 \end{bmatrix}$$

Computing $z_j - c_j$ for all of the nonbasic variables, we select the first row of the current $\hat{\mathbf{B}}_1^{-1}$, and by multiplying by the vectors for the nonbasic vectors, we have

$$(1, 1.9538, 0.2423, 1.3780)(\mathbf{a}_3^{(R)}, \mathbf{a}_5^{(R)}, \mathbf{a}_6^{(R)}, \mathbf{a}_7^{(R)})$$

$$= (1.4144, 1.9538, 0.2423, 1.3780)$$

Since no negatives are produced, we have an optimum solution and the final tableau is:

Final tableau

Variables in basis	β_0	β_1	β_2	β_3	$\mathbf{x}_B^{(R)}$	$\mathbf{y}_k^{(R)}$
z	1	1.9538	0.2423	1.3780	12737	O P
x_1	0	0.6471	−0.2353	0.1765	294.30	T I
x_4	0	0.5293	0.3529	−0.7646	58.80	M U
x_2	0	−0.5000	0.0000	0.5000	1500.00	M

A summary of these revised simplex tableaux is given in Sec. 6-11.

6-11 CONDENSED SUMMARY OF TABLEAUX

Variables in basis	β_0	β_1	β_2	β_3	$\mathbf{x}_B$	$\mathbf{y}_3^{(R)}$
z	1	0	0	0	000	−8.34
x_5	0	1	0	0	2000	2.40
x_6	0	0	1	0	8000	1.00
x_7	0	0	0	1	5000	3.50
						$\mathbf{y}_2^{(R)}$
z	1	3.4750	0	0	6950	−3.8250
x_3	0	0.4167	0	0	833.40	0.4167
x_6	0	−0.4167	1	0	7166.60	4.5833
x_7	0	−1.4583	0	1	2083.40	1.5417
						$\mathbf{y}_4^{(R)}$
z	1	−0.1430	0	2.4810	12119	−1.8420
x_3	0	0.8109	0	−0.2703	270.30	0.5406
x_6	0	3.9187	1	−2.9729	973.00	4.4458
x_2	0	−0.9459	0	0.6486	1351.20	−0.2973

Variables in basis	$\boldsymbol{\beta}_0$	$\boldsymbol{\beta}_1$	$\boldsymbol{\beta}_2$	$\boldsymbol{\beta}_3$	$\mathbf{x_B}$	$\mathbf{y}_3^{(R)}$
z	1	1.4809	0.4144	1.2490	12522	−0.7308
x_3	0	0.3344	−0.1216	0.0912	152.00	0.5168
x_4	0	0.8813	0.2249	−0.6686	218.80	0.5440
x_2	0	−0.6837	0.0669	0.4497	1416.30	−0.2841
						$\mathbf{y}_k$
z	1	1.9538	0.2423	1.3780	12737	O P
x_1	0	0.6471	−0.2353	0.1765	294.30	T I
x_4	0	0.5293	0.3529	−0.7646	58.80	M U
x_2	0	−0.5000	0.0000	0.5000	1500.00	M

6-12 SUMMARY OF REVISED SIMPLEX METHOD

The problem above has been carried out in great detail in order to make the process clear. It will be noted that a complete tableau was not given at each iteration but rather only the inverse of the basis set of vectors, together with the solutions vector and the entering vector. With this information and the initial tableau as given on p. 131, everything was developed as needed. The elements of the Summary of Sec. 6-11 can, of course, be found on p. 70 but not in the same form. The columns here denoted by $\boldsymbol{\beta}_i$ and $\mathbf{y}_i^{(R)}$ were denoted by the different x_i's throughout the tableaux.

In Summary 6-11 it would not have been necessary to carry the column corresponding to the z variable, since it remains as a unit vector throughout the computation. The portion enclosed in the rectangle for each of the iterations is the inverse of the basis set of vectors as carried out in the regular simplex procedure and the single row on the line designated with z is the $z_j - c_j$ row.

While there are both advantages and disadvantages in using the revised simplex method, it is included here for completeness. Generally speaking, one would only use it in a computer routine, and many computer codes offer this as an option to the user. In large-scale problems, less memory is usually required. The method has particular advantages when solving problems by the decomposition method. Further reference is made to this method on p. 208.

EXERCISES

The exercises for this chapter are mostly ones of concept rather than computation. The MPOS program to which reference has been made throughout the text offers the option of using the revised simplex method of solution if one chooses to do so.

1 Discuss why you think the revised simplex method might not be used for problems done by hand-computational methods.

2 Make an estimate of the number of computations required to move from one iteration to the next in the simplex method.

3 Make an estimate of the number of computations required to move from one iteration to the next in the revised simplex method.

4 In using the revised simplex method, the entire $z_j - c_j$ row is normally computed, and the largest negative term is selected in order to identify the incoming vector. In this method, would it be possible and also convenient to select the first negative $z_j - c_j$ and use that to determine the next incoming vector? Are there any advantages?

5 Are there any advantages in terms of memory required for the revised simplex method versus the regular simplex method?

CHAPTER

SEVEN

THE TRANSPORTATION MODEL

7-1 INTRODUCTION TO THE TRANSPORTATION MODEL

A certain class of linear programming problems, known as transportation-type problems, arises frequently. The properties that it possesses enable one to solve the problem by methods that are considerably more efficient than either the regular or the revised simplex methods. The problem has its origin in devising a schedule for shipping products from m origins to n destinations, where the unit cost of shipping the product from any origin to any destination is known. Furthermore, the quantity available at each origin and the quantity required at each destination is known.

To express the problem mathematically, let

x_{ij} = amount to be shipped from origin i to destination j
c_{ij} = cost of shipping one unit from origin i to destination j
a_i = amount available at origin i
b_j = amount required at destination j

The problem whose solution is required is then
Find

$$x_{ij} \geqq 0$$

such that

$$\sum_{j=1}^{n} x_{ij} = a_i \qquad i = 1, 2, 3, \ldots, m$$

$$\sum_{i=1}^{m} x_{ij} = b_j \qquad j = 1, 2, 3, \ldots, n$$

and such that

$$Z = \sum_{i=1}^{m} \sum_{j=1}^{n} c_{ij} x_{ij} \text{ is a } \textit{minimum}.$$

There is an added inherent restriction that the amount of product available is equal to the amount that is required and it is expressed by

$$\sum_{i=1}^{m} a_i = \sum_{j=1}^{n} b_j$$

This last restriction will later be removed.

7-2 SPECIAL FORM OF PROBLEM—DUAL STATEMENT

If one were to solve this problem by either the regular or revised simplex method, one would have to add an artificial variable to each of the constraints so that the resulting problem would then consist of $m + n$ constraints in $m \times n + m + n$ variables. Even for relatively small m and n, the problem becomes unwieldy and essentially impossible to handle by manual methods, whereas the special algorithms enable one to handle rather large problems with relative ease. The special approach is dependent upon writing the dual statement of the problem and developing an algorithm for the dual problem, which in turn is converted back to the solution for the primal problem. To avoid some of the difficulties of the general notation, we shall develop the ideas by use of a problem that has only three origins and four destinations. Specifically, let us consider the problem represented by the tableau

	D_1	D_2	D_3	D_4	Amounts Available
O_1	x_{11} [12]	x_{12} [3]	x_{13} [8]	x_{14} [4]	$7 = a_1$
O_2	x_{21} [7]	x_{22} [4]	x_{23} [6]	x_{24} [9]	$10 = a_2$
O_3	x_{31} [8]	x_{32} [7]	x_{33} [3]	x_{34} [6]	$12 = a_3$
Amounts Required	$4 = b_1$	$8 = b_2$	$11 = b_3$	$6 = b_4$	

in which i varies from 1 to 3 and j varies from 1 to 4. It can be seen for this example that

$$\sum_{i=1}^{3} a_i = \sum_{j=1}^{4} b_j = 29$$

In this format we have all the information necessary for the solution of the problem. One obvious solution to this problem is

$$x_{11} = 4 \qquad x_{12} = 3 \qquad x_{22} = 5 \qquad x_{23} = 5 \qquad x_{33} = 6 \qquad x_{34} = 6$$

with a cost of 161. If these values are placed within the transportation tableau as shown above, we have

	D_1	D_2	D_3	D_4	Amounts available
O_1	(4) 12	(3) 3	8	4	7
O_2	7	(5) 4	(5) 6	9	10
O_3	8	7	(6) 3	(6) 6	12
Amounts required	4	8	11	6	

The circled number in each cell is the amount assigned and the small number in the upper right-hand corner of the cell is the unit cost. For this problem the standard form of the statement of the problem is:

Find

$$x_{ij} \geqq 0$$

such that

$$\begin{array}{llllllllllllr}
x_{11} + x_{12} + x_{13} + x_{14} & & & & & & & & & & & & = 7 \\
 & & & & x_{21} + x_{22} + x_{23} + x_{24} & & & & & & & & = 10 \\
 & & & & & & & & x_{31} + x_{32} + x_{33} + x_{34} & & & & = 12 \\
x_{11} & & & + x_{21} & & & & + x_{31} & & & & & = 4 \\
 & x_{12} & & & + x_{22} & & & & + x_{32} & & & & = 8 \\
 & & x_{13} & & & + x_{23} & & & & + x_{33} & & & = 11 \\
 & & & x_{14} & & & + x_{24} & & & & + x_{34} & & = 6
\end{array}$$

and such that

$12x_{11} + 3x_{12} + 8x_{13} + 4x_{14} + 7x_{21} + 4x_{22} + 6x_{23} + 9x_{24} + 8x_{31} + 7x_{32} + 3x_{33} + 6x_{34}$ is a *minimum.*

If we define the dual variables u_1, u_2, and u_3 as corresponding to the first three constraints and v_1, v_2, v_3, and v_4 as corresponding to the last four constraints, the dual statement of the problem becomes to find u_i and v_j, each unrestricted in sign, so that

$$u_i + v_j \leqq c_{ij}$$

such that

$$7u_1 + 10u_2 + 12u_3 + 4v_1 + 8v_2 + 11v_3 + 6v_4$$

is a maximum. The general statement of the objective function to be maximized is

$$\sum_{i=1}^{3} a_i u_i + \sum_{j=1}^{4} b_j v_j$$

Fortunately, this dual problem is an easier problem to solve than the primal problem. It so happens that it is possible to solve for u_i and v_j quite easily by taking advantage of the structure of the problem.

In any $m \times n$ transportation model, a basic feasible solution (nondegenerate) will have $m + n - 1$ variables with nonzero values. This will always be true, since the primal problem has $m + n$ constraints, one of which is dependent upon the others, since it is required that

$$\sum_{i=1}^{3} a_i = \sum_{j=1}^{4} b_j$$

Thus we have $m + n - 1$ independent constraints and hence the same number of basic solutions.

7-3 DEVELOPMENT OF SOLUTION ROUTINE

With this in mind, let us develop the solution for the example problem. The sequence of steps is as follows:

1. Arrange the problem in an $m \times n$ tableau with the row requirements on the right and the column requirements at the bottom. Each cell should contain the unit cost appropriate for the shipment.
2. Obtain a first feasible solution by making assignments in each of the cells so as to meet the requirements. Any method may be used, but when a cell is selected for an assignment, the maximum amount possible must be assigned in order to have a basic solution. Let us use and illustrate the *northwest corner* rule. Select cell $O_1 D_1$; the maximum amount will be the lesser of the row requirement and the column requirement, which in this case is 4. This will complete the first column requirement and reduce the first row requirement to 3. We now have a 3×3 tableau to be satisfied. Move to cell $O_1 D_2$; assign an amount that is the smaller of the column requirement of 8 and the remaining row requirement of 3. The amount is 3, which is assigned to $O_1 D_2$. The first row requirement is now satisfied and the second column requirement is reduced to 5. This process is continued until all row and column requirements are satisfied. The method is not particularly efficient, so better methods will be discussed later. It will be found that when the last assignment is made, the final row and column requirements are satisfied simultaneously. When one starts with m rows and n columns, the maximum number of assignments will be

$m + n - 1$. If there are fewer than $m + n - 1$ assignments, we have a degenerate solution. Consider first the nondegenerate case.

3. Write the dual statements corresponding to the *assigned cells*. Write the constraints as equations. In our example we have

$$u_1 + v_1 = 12$$
$$u_1 + v_2 = 3$$
$$u_2 + v_2 = 4$$
$$u_2 + v_3 = 6$$
$$u_3 + v_3 = 3$$
$$u_3 + v_4 = 6$$

It will be noted that we have six equations in seven unknowns, which in general leads to an infinite number of solutions; that is, one of the variables can be assigned an arbitrary value, reducing the system to six equations in six unknowns. For our example let us assign $u_1 = 0$. Then we have $v_1 = 12$ and $v_2 = 3$. Using $v_2 = 3$, we find $u_2 = 1$. Using $u_2 = 1$, we find $v_3 = 5$. Using $v_3 = 5$, we find $u_3 = -2$, and finally using $u_3 = -2$, we find $v_4 = 8$. Thus the dual constraints corresponding to each of the occupied cells are satisfied.

4. We must now test the constraints corresponding to each of the unoccupied cells for the values of u_i and v_j given above. Schematically, we can write these values as follows:

Initial tableau

	$v_1 = 12$	$v_2 = 3$	$v_3 = 5$	$v_4 = 8$	
$u_1 = 0$	(4) [12]	(3) [3]	[8]	[4]	7
$u_2 = 1$	[7]	(5) [4]	(5) [6]	[9]	10
$u_3 = -2$	[8]	[7]	(6) [3]	(6) [6]	12
	4	8	11	6	

For cell O_1D_3, $u_1 + v_3 \leqq 8$ is satisfied

O_1D_4, $u_1 + v_4 \leqq 4$ is *not* satisfied

O_2D_1, $u_2 + v_1 \leqq 7$ is *not* satisfied

O_2D_4, $u_2 + v_4 \leqq 9$ is satisfied

O_3D_1, $u_3 + v_1 \leqq 8$ is *not* satisfied

O_3D_2, $u_3 + v_2 \leqq 7$ is satisfied

Thus it can be seen that three of these constraints are *not* satisfied for this solution (feasible) to the primal, and thus it cannot be optimal. These correspond to cells $O_1 D_4$, $O_2 D_1$, and $O_3 D_1$. Let us examine the cell $O_2 D_1$ and adjacent cells shown below. These same cells are outlined with heavy lines in the tableau above, and the isolated section is shown below.

	D_1 $V_1 = 12$	D_2 $v_2 = 3$
O_1 $u_1 = 0$	(4) [12]	(3) [3]
O_2 $u_2 = 1$	[7]	(5) [4]

It can be seen that 5 units can be moved from $O_2 D_2$ to $O_2 D_1$ at an increase of 3 cost units for each unit moved. Similarly, 4 units can be moved from $O_1 D_1$ to $O_1 D_2$ at a decrease of 9 cost units for each unit moved. We can make both of these moves simultaneously at a net decrease of 6 units of cost for each unit moved. We must, however, move equal amounts for the two moves in order to maintain a total of 4 units in column D_1 and 8 units in column D_2. The row requirements will not be altered since the movement of units is made along rows. It will be noted that this net reduction per unit moved is equal to

$$c_{21} - (u_2 + v_1)$$

which can be shown as follows:

Adding 1 unit to cell $O_2 D_1$ adds a cost of c_{21}.
Subtracting 1 unit from cell $O_2 D_2$ reduces cost c_{22}.
Adding 1 unit to cell $O_1 D_2$ adds a cost of c_{12}.
Subtracting 1 unit from cell $O_1 D_1$ reduces cost c_{11}.

Thus the net change in cost per unit moved is

$$c_{21} - c_{22} + c_{12} - c_{11}$$

But since

$$u_2 + v_2 = c_{22}$$

$$u_1 + v_2 = c_{12}$$

$$u_1 + v_1 = c_{11}$$

by substitution, the net change is

$$c_{21} - (u_2 + v_2) + (u_1 + v_2) - (u_1 + v_1) = c_{21} - u_2 - v_1$$

Thus if

$$u_2 + v_1 > c_{21}$$

a reduction can be made in the amount of

$$u_2 + v_1 - c_{21}$$

per unit transferred. For our example

$$u_2 + v_1 - c_{21} = 1 + 12 - 7 = 6$$

The maximum number of units that can be moved at this rate is 4 units. Therefore, the reduction in cost over the first feasible solution is 6 times 4, or 24 units, if these transfers are made. To examine cell $O_3 D_1$, let us review the cell together with its adjacent cells as shown below.

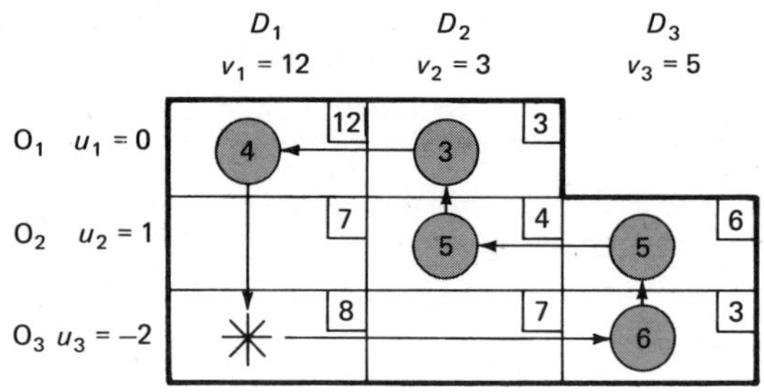

The way to easily assess the transfers that can be made is to start with the cell being examined and move horizontally until an occupied cell is reached, then move vertically to an occupied cell, then horizontally and vertically until you are able to return to the cell being examined. At times there may be more than one occupied cell in either a horizontal or a vertical move. If that is the case, a choice must be made so that it is possible to return to the original cell. It can be shown that the path to be traced is unique. The unique path is shown above for the $O_3 D_1$ cell. For this path, the costs are as follows:

$$8 - 3 + 6 - 4 + 3 - 12 = -2$$

This is a reduction of 2 units of cost per unit transferred. Since the first of the two moves considered above is more advantageous, that move will be made and the result is given below. With this change we must find a new set of u_i's and v_j's, again starting with $u_1 = 0$.

Second tableau

(Each cell shows the cost c_{ij} in brackets, followed by the allocation where one is made.)

	$v_1 = 6$	$v_2 = 3$	$v_3 = 5$	$v_4 = 8$	
$u_1 = 0$	[12]	[3] 7	[8]	[4]	7
$u_2 = 1$	[7] 4	[4] 1	[6] 5	[9]	10
$u_3 = -2$	[8]	[7]	[3] 6	[6] 6	12
	4	8	11	6	

We now see that the only cell for which $u_i + v_j$ exceeds c_{ij} is $O_1 D_4$, and therefore the maximum amount possible should be moved into that cell. Five units can be moved into that cell so that, along with some other transfers, the resulting tableau is:

Third tableau

	$v_1 = 6$	$v_2 = 3$	$v_3 = 1$	$v_4 = 4$	
$u_1 = 0$	[12]	[3] 2	[8]	[4] 5	7
$u_2 = 1$	[7] 4	[4] 6	[6]	[9]	10
$u_3 = 2$	[8]	[7]	[3] 11	[6] 1	12
	4	8	11	6	

This tableau is now optimal, since there is no cell for which $u_i + v_j$ exceeds c_{ij}. It can be seen that $u_3 + v_1 = c_{31}$, indicating that a transfer may be made into that cell without changing the value of the objective function. Hence an alternate solution exists.

The example given shows the steps that must be taken to solve the transportation problem. The illustration was chosen so that the feasible solutions at each stage contained no degeneracies. Degeneracies can be handled by modifications to the above as shown in Sec. 7.4.

The concepts on transportation problems can be generalized to a more comprehensive class of network flow problems. Minimal-cost network flow problems occur in the design and analysis of communication systems, oil pipeline systems, tanker scheduling problems, and a variety of other areas. The minimal cost flow problem is a linear program. In fact, transportation problems are a special case of network flows.

The following flow chart shows in a simplified form the steps that must be taken to solve the transportation model.

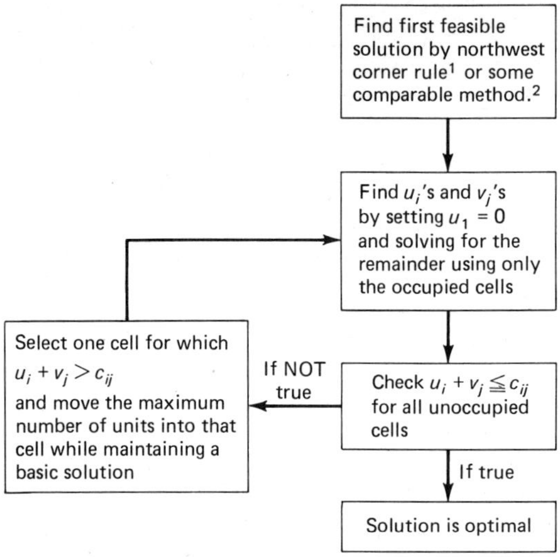

[1] See Sec. 7-3.

[2] There are many methods of getting a first feasible solution, one of which is Vogel's approximation method. Even experience and intuition might be used.

7-4 DEGENERACIES IN THE SOLUTION

One or more degeneracies may cause some slight complications in the procedure described above. To illustrate the situation, consider a slight alteration in the example we have been using. If we have a first-row availability of 8 instead of 7, and a first-column requirement of 5 instead of 4 and again using the *northwest corner* rule, we will have the first tableau shown below.

Initial tableau

	$v_1 = 12$	$v_2 = 3$	$v_3 = 5$	$v_4 = 8$	
$u_1 = 0$	(5) 12	(3) 3	5	8	8
$u_2 = 1$	7	(5) 4	(5) 6	9	10
$u_3 = -2$	8	7	(6) 3	(6) 6	12
	5	8	11	6	

This initial tableau, together with its dual variables, has been obtained by the procedure previously described. By examining the unoccupied cells, it is seen that

$$u_i + v_j \text{ exceeds } c_{ij}$$

by the greatest amount when $i = 2$ and $j = 1$, so that some units will be transferred into cell $O_2 D_1$. Five units will be transferred from $O_2 D_2$ to $O_2 D_1$ and a corresponding transfer of 5 units from $O_1 D_1$ to $O_1 D_2$. The resulting tableau is shown next.

Second tableau

(Each cell: allocation in circle, if any; cost c_{ij} in box.)

	$v_1 = 6$	$v_2 = 3$	$v_3 = 5$	$v_4 = 8$	
$u_1 = 0$	[12]	(8) [3]	[5]	[8]	8
$u_2 = 1$	(5) [7]	(0) [4]	(5) [6]	[9]	10
$u_3 = -2$	[8]	[7]	(6) [3]	(6) [6]	12
	5	8	11	6	

It should be noted that cell $O_2 D_2$ contains "zero" units. Had this not been entered, there would have been only five occupied cells. This is a degenerate solution since it has less than $m + n - 1$ assignments. This makes it impossible to calculate the dual variables. After setting $u_1 = 0$, $v_2 = 3$ but no further solutions can be found. It is necessary, therefore, to artificially supply an occupied cell either in row one or in column two. This will then provide the connecting link for calculating the remainder of the dual variables (cell $O_2 D_2$ was arbitrarily selected). It is found by testing all unoccupied cells that the solution is optimal, since no $u_i + v_j > c_{ij}$. As mentioned, the solution is degenerate since one of the basic solutions is equal to zero. It will be further noted that alternate optimal solutions exist. Five units can be transferred into cell $O_1 D_4$. The result is:

Alternate optimal tableau

	$v_1 = 6$	$v_2 = 3$	$v_3 = 5$	$v_4 = 8$	
$u_1 = 0$	[12]	(3) [3]	[5]	(5) [8]	8
$u_2 = 1$	(5) [7]	(5) [1]	[6]	[9]	10
$u_3 = -2$	[8]	[7]	(11) [3]	(1) [6]	12
	5	8	11	6	

The alternate optimal solution is no longer degenerate. The degeneracies occur here in a manner similar to that for linear programming problems in general. The degeneracies are handled much more easily in the transportation model.

The method demonstrated above can be used to solve any transportation-type problem. Because of the simplicity of the model, the solution lends itself to manual computation. Although it will rarely be necessary to do so, it is possible to solve fairly large problems manually within a short period of time.

7-5 GENERALIZING THE TRANSPORTATION MODEL

On p. 145, the requirement that

$$\sum_{i=1}^{3} a_i = \sum_{j=1}^{4} b_j$$

was imposed. This means that the sum of the requirements is exactly equal to the sum of the availabilities. Such a requirement was imposed purely for convenience in developing the method of solution. Suppose that in our example on p. 145, the amounts available at the origins were 8, 11, and 14 instead of the amounts originally stated. Thus the total amount available is 33 units, and the amount required at the four destinations is only 29 units. To place this problem in the same model format as in the demonstration example, we create a fictitious destination, D_5. The variables corresponding to destination D_5 are really slack variables, and the unit shipping costs from each of the origins must be equal in order not to show preference. We have assigned zeros for simplicity. The resulting tableau will then be:

	D_1	D_2	D_3	D_4	D_5	Amounts available
O_1	(4) 12	(4) 3	8	4	0	8
O_2	7	(4) 4	(7) 6	9	0	11
O_3	8	7	(4) 3	(6) 6	(4) 0	14
Amounts required	4	8	11	6	4	

with the first feasible solution shown above. Using the regular method of solving this problem, the resulting optimal tableau is:

	D_1	D_2	D_3	D_4	D_5	Amounts available
O_1	12	(2) 3	8	(6) 4	0	8
O_2	(4) 7	(6) 4	6	9	(1) 0	11
O_3	8	7	(11) 3	6	(3) 0	14
Amounts required	4	8	11	6	4	

The amounts assigned to D_5 could be interpreted to mean that these items were not shipped at all. Thus 1 unit available at O_2 and 3 units available at O_3 were not shipped at all. The rest of the assignments, of course, have the ordinary interpretation. We refer to D_5 as a dummy destination and such will be used when the amount available exceeds the amount required. If the reverse is true, a dummy row is added to the tableau.

7-6 EFFICIENT FIRST FEASIBLE SELECTION

In Sec. 7-3 it was indicated that the *northwest corner* rule was not a particularly efficient method for obtaining a first feasible solution. If one has an intuitive method of getting a first feasible solution, by all means it should be used. One method used is to select for the first assignment a cell that takes into account any penalty that might be incurred if one is forced to take the second choice. For example, in our sample problem on p. 145, if one were to select from the first row, he or she would choose the cell with a cost of 3. However, if one is unable to select the 3 and is forced to take the second best choice, it would have a cost of 4. Thus we might say that the penalty in that row is 1, the difference between the second best choice and the best choice. Similarly, we have the following penalties for the entire choices of all rows and columns:

First row	1
Second row	2
Third row	3
First column	1
Second column	1
Third column	3
Fourth column	2

From this review it is seen that one would prefer not to have to take the second choice in either the third row or the third column, and as a result could make the first choice in either one or the other of these cases. It so happens that the first choice for each of these cases results in an assignment to cell $O_3 D_3$, and when the maximum amount is computed, we find that 11 units will be assigned to that cell, with the following resulting tableau:

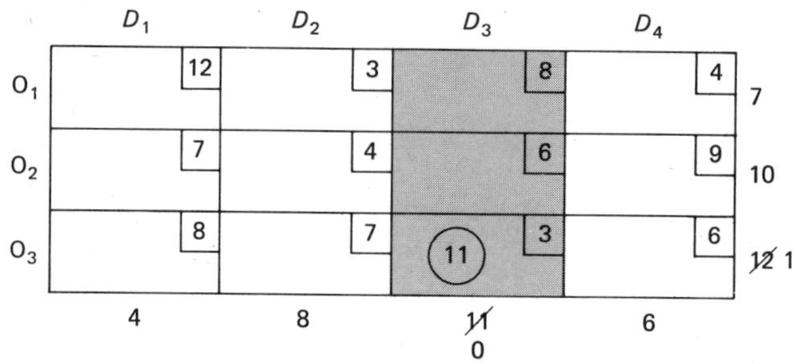

The requirements for column three have now been completed and the requirements for row three have been reduced to 1. Consider the reduced tableau obtained by removing from consideration the column three that has been shaded out. The penalties are then:

Row one	1
Row two	3
Row three	1
Column one	1
Column two	1
Column four	2

Because of the penalty of 3 for the second row, we should select the best in the second row. This is cell $O_2 D_2$, and 8 units can be assigned, resulting in the following tableau.

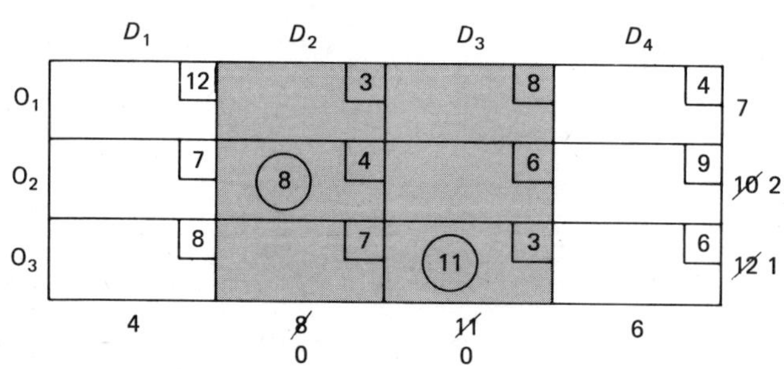

The requirements for columns two and three have been completed and those for rows two and three have been reduced. Consider the reduced tableau obtained by removing from consideration the columns two and three that have been shaded out. The penalties are then:

Row one	8
Row two	2
Row three	2
Column one	1
Column four	2

Because of the penalty of 8 for the first row, we should select the best in the first row. This is cell $O_1 D_4$ and 6 units can be assigned, resulting in the following tableau:

	D_1	D_2	D_3	D_4	
O_1	12	3	8	(6) 4	~~7~~ 1
O_2	7	(8) 4	6	9	~~10~~ 2
O_3	8	7	(11) 3	6	~~12~~ 1
	4	~~8~~ 0	~~11~~ 0	~~6~~ 0	

The only possible assignments remaining are now in the first column, and the amounts to be assigned are the remaining requirements to be satisfied as shown. The resulting first feasible tableau is then shown, together with the calculated values of the dual variables.

	$v_1 = 12$	$v_2 = 9$	$v_3 = 7$	$v_4 = 4$	
$u_1 = 0$	(1) 12	3	8	(6) 4	7
$u_2 = -5$	(2) 7	(8) 4	6	9	10
$u_3 = -4$	(1) 8	7	(11) 3	6	12
	4	8	11	6	

Checking the unoccupied cells it is found that the only cell for which $u_i + v_j$ exceeds the corresponding c_{ij} is cell $O_1 D_2$, for which a per unit improvement of 6 can be obtained for each unit moved into that cell. Only 1 unit can be moved, resulting in the optimal tableau shown next.

	$v_1 = 6$	$v_2 = 3$	$v_3 = 1$	$v_4 = 4$	
$u_1 = 0$	12	(1) 3	8	(6) 4	7
$u_2 = 1$	(3) 7	(7) 4	6	9	10
$u_3 = 2$	(1) 8	7	(11) 3	6	12
	4	8	11	6	

This optimal tableau is not the same as that shown on p. 151—it is an alternate. The dual variables are the same, however. In the current tableau, the cell $O_3 D_4$ can be used for an assignment without changing the value of the objective function. Using that cell will result in the same alternate optimal as that shown on p. 151.

The process described above for obtaining a first feasible solution is one of many that can be used to get a good starting point. It is very efficient even in very large problems and can be carried out manually quite easily. The first feasible solution must, of course, be basic. Knowledge of the structure of the problem may help in obtaining a good first feasible solution, although the choice does not affect the optimality of the final result. A poor choice may lead to a number of additional iterations, but this is not itself much of a problem if a computer is used for the solution, unless time is a problem.

One must weigh the costs of spending the time on developing an efficient first feasible solution versus the cost of the additional computer time. If a problem is to be solved on a repetitive basis, it becomes more important to develop a highly efficient basic first feasible solution.

7-7 ASSIGNMENT PROBLEMS

The discussion in this chapter was devoted to the transportation model. Since it is a special case of a linear programming problem, it can of course be solved by either the regular simplex method or the revised simplex method at the cost of some computing time. Although it will not be discussed here, a problem known as the assignment problem is, in turn, a special case of the transportation model. An algorithm that is highly efficient exists for its solution.

EXERCISES

The exercises in this chapter are modest in nature, since each of these problems can always be solved by the simplex method. The chapter and the exercises are given to round out the linear programming ideas and to give some examples that might be of interest to those concerned with distribution problems.

1 Consider the following problem in shipping cases of detergent from designated plants to warehouses. Availabilities from the plants are listed on the right and requirements of the warehouses are given at the bottom of each of the appropriate columns. The per unit shipping costs are given in the body of the table.

Plant	Warehouse						Available quantities
	Peoria	Elgin	Rockford	Genoa	Forest	Lake	
Misha	1	1	5	2	4	2	30
Wauka	3	2	4	3	1	3	50
Osh	4	5	6	2	9	2	75
Kosh	3	7	4	6	3	1	20
Requirements	20	30	50	25	10	40	

Find the way in which the available quantities should be shipped to the designated warehouses to minimize the total shipping costs. Try your hand at finding a first feasible solution just by examining the problem carefully. You need not use any of the methods discussed.

2 For the same plants and warehouses as in Exercise 1, consider the distribution of cases of salad oil. Requirements, availabilities, and per unit shipping costs are given again.

Plant	Warehouse						Available quantities
	Peoria	Elgin	Rockford	Genoa	Forest	Lake	
Misha	5	15	9	7	8	10	30
Wauka	14	10	20	21	9	13	40
Osh	15	13	8	12	25	11	10
Kosh	9	12	6	13	8	19	100
Requirements	50	10	15	50	35	20	

Find the optimal shipping schedule to minimize total shipping costs.

3 A processor and canner of tuna has three plants from which to ship to six major distribution points. The table shows the shipping costs per case as well as the predicted demands for the coming year in millions of cases. The variable production costs per case are: Seattle, 0.85; Portland, 0.72; Tacoma, 0.94.

Determine the minimum-cost production-shipping schedule for the coming year on the basis of predicted demands.

Plant	Distribution point						Available quantities
	Vegas	Reno	Taho	Evanston	Wahoo	Marin	
Seattle	0.26	0.54	0.20	0.37	0.41	0.30	2.2
Portland	0.53	0.41	0.35	0.25	0.20	0.40	1.8
Tacoma	0.45	0.30	0.32	0.41	0.50	0.35	3.4
Requirements	0.85	0.42	1.02	0.92	0.58	0.75	

4 Show that if the variable production costs are exactly the same at each of the plants in Seattle, Portland, and Tacoma, the optimal production-shipping schedule can be determined by using shipping costs only.

5 Prove that if every one of the processing plants must be operated at capacity, the optimal shipping schedule can be obtained by using transportation costs only, even though the production costs differ from plant to plant.

CHAPTER

EIGHT

COMPUTER-GENERATED INFORMATION

8-1 USE OF THE COMPUTER

It must be evident from the material covered in the preceding chapters that the computer is essential for solving linear programming problems except for the very simple ones. To intelligently discuss the use of the computer, it is necessary to relate the discussion to some particular "canned" program. The discussion that follows is centered around MPOS†, developed at Northwestern University. The purpose of the discussion is to give some insight into the ease with which a canned program can be used and the wealth of information that can be obtained from a well-planned solution.

8-2 SCHEDULING PROBLEM

Consider the following scheduling problem. A company manufactures three products, *A*, *B*, and *C*. The maximum sales potential for these products is 30,000; 120,000; and 150,000 units per month, respectively. The company must manufacture at least 10,000; 20,000; and 10,000 units, respectively, each month in order to meet contractual agreements. The contributions to profit (exclusive of manufacturing costs) are 75, 15, and 10 cents/unit, respectively.

† Multi-Purpose Optimization System—*User's Guide* by Claude Cohen and Jack Stein, Vogelback Computing Center, Northwestern University, Evanston, Illinois 60201.

Each unit must go through four manufacturing stages. The production capacity is presently 40, 250, 75, and 30 h/shift for stages 1, 2, 3, and 4, respectively. There are 20 working days, 2 shifts/day, for the month. The number of production hours required to process 1000 units of each product through each stage is given in Table 8-1.

Table 8-1 Production hours

	Product		
Stage	*A*	*B*	*C*
1	24	40	20
2	600	70	40
3	180	10	15
4	9	30	25

The variable cost per hour of operating the stages day and night is given in Table 8-2.

Table 8-2 cost per hour

Stage	Day	Night
1	0.10	0.30
2	0.50	0.60
3	0.30	0.50
4	0.90	0.90

On the assumption that sequencing is not a problem, find, by product, by stage, and by shift, the allocation of production that will maximize profit for the month. We eliminate the sequencing problem by having enough in process inventory at various stages of completion so that no one ever has to wait for work.

To formulate this problem, first let the amounts to be produced be measured in units of 1000. Then let

x_1 = number of units of *A* to be produced
x_2 = number of units of *B* to be produced
x_3 = number of units of *C* to be produced

The following table gives the assignment of variables which indicates the amount of each variable that is assigned to each stage for the day and for the night.

	Day			Night		
Stage	*A*	*B*	*C*	*A*	*B*	*C*
1	x_4	x_8	x_{12}	x_{16}	x_{20}	x_{24}
2	x_5	x_9	x_{13}	x_{17}	x_{21}	x_{25}
3	x_6	x_{10}	x_{14}	x_{18}	x_{22}	x_{26}
4	x_7	x_{11}	x_{15}	x_{19}	x_{23}	x_{27}

From sales and contractual agreements, the constraints on quantities are:

$$x_1 \leqq 30$$

$$x_2 \leqq 120$$

$$x_3 \leqq 150$$

$$x_1 \geqq 10$$

$$x_2 \geqq 20$$

$$x_3 \geqq 10$$

It is seen from the variable definition that a total of $x_4 + x_{16}$ units of *A* in stage 1 is produced in the day and night together. Similarly, a total of $x_5 + x_{17}$ units of *A* in stage 2 is produced in the day and night together; and so on. In order that we produce the same number of each product for each stage, we must have:

Stage balances for *A*:

$$x_1 = x_4 + x_{16}$$

$$x_1 = x_5 + x_{17}$$

$$x_1 = x_6 + x_{18}$$

$$x_1 = x_7 + x_{19}$$

Stage balances for *B*:

$$x_2 = x_8 + x_{20}$$

$$x_2 = x_9 + x_{21}$$

$$x_2 = x_{10} + x_{22}$$

$$x_2 = x_{11} + x_{23}$$

Stage balances for *C*:

$$x_3 = x_{12} + x_{24}$$

$$x_3 = x_{13} + x_{25}$$

$$x_3 = x_{14} + x_{26}$$

$$x_3 = x_{15} + x_{27}$$

Additional constraints are required in order to stay within the production capacity for each stage for each shift. For example, for stage 1,

$$24x_4 + 40x_8 + 20x_{12}$$

is the number of hours used, and since we have 800 h available in stage 1 during the day, we must have

$$24x_1 + 40x_8 + 20x_{12} \leqq 800$$

The entire collection of stage constraints is as follows:

Stage constraints day:

$$\begin{aligned} 24x_4 + 40x_8 + 20x_{12} &\leqq 800 \\ 600x_5 + 70x_9 + 40x_{13} &\leqq 5000 \\ 180x_6 + 10x_{10} + 15x_{14} &\leqq 1500 \\ 9x_7 + 30x_{11} + 25x_{15} &\leqq 600 \end{aligned}$$

Stage constraints night:

$$\begin{aligned} 24x_{16} + 40x_{20} + 20x_{24} &\leqq 800 \\ 600x_{17} + 70x_{21} + 40x_{25} &\leqq 5000 \\ 180x_{18} + 10x_{22} + 15x_{26} &\leqq 1500 \\ 9x_{19} + 30x_{23} + 25x_{27} &\leqq 600 \end{aligned}$$

Thus we have 26 constraints with 27 variables, exclusive of the slacks and artificials that will be added by the program.

The profit consists of income less variable cost. The income can be conveniently assigned to the variables x_1, x_2, and x_3 and the production costs to variables x_4 through x_{27}. With these assignments, we have:

$$\text{income} = 750x_1 + 150x_2 + 100x_3$$

The costs for each stage for each shift are obtained by multiplying costs per hour by hours required for each product by shift and stage. Thus the costs will be

$$\begin{aligned} &2.4x_4 + 300x_5 + 54x_6 + 8.1x_7 \\ &+ 4x_8 + 35x_9 + 3x_{10} + 27x_{11} \\ &+ 2x_{12} + 20x_{13} + 4.5x_{14} + 22.5x_{15} \\ &+ 7.2x_{16} + 360x_{17} + 90x_{18} + 8.1x_{19} \\ &+ 12x_{20} + 42x_{21} + 5x_{22} + 27x_{23} \\ &+ 6x_{24} + 24x_{25} + 7.5x_{26} + 22.5x_{27} \end{aligned}$$

The contribution to profit is then given by income less cost. With this complete specification of the problem, we will now indicate the exact manner in which it must be specified in order to get the solution by means of the MPOS program at Northwestern University. The program has been designed to accept algebraic statements such as those given above. Exhibit 8-1 shows the form of the input, line by line, exactly as it was entered. The objective function was entered first, together with the indication for maximization. The constraints were then entered, interspersed with comments, each of which was preceded by an asterisk. The program supplied the numbering of the constraints for easy identification purposes in the later analyses. All constraints were entered with constants only, to the right of the inequality or equals signs. The right-hand sides are nonnegative.

Exhibit 8-1

```
***** PROBLEM NUMBER  1 *****

        *L-TRONIC LARGE MODEL
        VARIABLES
        X1 TO X27
        MAXIMIZE
        750X1+150X2+100X3-2.4X4-300X5-54X6-8.1X7-4X8-35X9-3X10
        -27X11-2X12-20X13-4.5X14-22.5X15-7.2X16-360X17-90X18-8.1X19-12X20
        -42X21-5X22-27X23-6X24-24X25-7.5X26-22.5X27
        CONSTRAINTS
        *SALES CONSTRAINTS
   1.   X1.LE.30
   2.   X2.LE.120
   3.   X3.LE.150
   4.   X1.GE.10
   5.   X2.GE.20
   6.   X3.GE.10
        *BALANCE STAGES FOR A
   7.   X1-X4-X16=0
   8.   X1-X5-X17=0
   9.   X1-X6-X18=0
  10.   X1-X7-X19=0
        *BALANCE STAGES FOR B
  11.   X2-X8-X20=0
  12.   X2-X9-X21=0
  13.   X2-X10-X22=0
  14.   X2-X11-X23=0
        *BALANCE STAGES FOR C
  15.   X3-X12-X24=0
  16.   X3-X13-X25=0
  17.   X3-X14-X26=0
  18.   X3-X15-X27=0
        *CAPACITY CONSTRAINTS FOR THE DAY
  19.   24X4+40X8+20X12.LE.800
  20.   600X5+70X9+40X13.LE.5000
  21.   180X6+10X10+15X14.LE.1500
  22.   9X7+30X11+25X15.LE.600
        *CAPACITY CONSTRAINTS FOR THE NIGHT
  23.   24X16+40X20+20X24.LE.800
  24.   600X17+70X21+40X25.LE.5000
  25.   180X18+10X22+15X26.LE.1500
  26.   9X19+30X23+25X27.LE.600
        PRINT
        RNGOBJ
        RNGRHS
        OPTIMIZE
```

No slacks, surpluses, or artificials have been entered—these are generated as needed by the program and are later shown in the output.

Following the constraints, four instructions are given:

PRINT. This calls for a printing of the initial and the final tableau.
RNGOBJ. This calls for pertinent information that enables one to make a sensitivity analysis of the coefficients of the objective function.
RNGRHS. This calls for pertinent information that enables one to do a sensitivity analysis on the right-hand sides of the constraints.
OPTIMIZE. This is the signal to go ahead and solve the stated problem.

8-3 SOLUTION—NATURE OF THE RESULTS

In order to get the most out of a solution to a linear programming problem, one must get a number of results. The following are obtained from this program:

1. The actual solution to the problem, including values for all original variables as well as those which have been added to get a solution to the problem. These are given under the heading "Summary of Results."
2. Specific indication of which constraints are binding and which are nonbinding. These results are also given under the heading "Summary of Results."
3. Sensitivity analysis for the coefficients of the objective function. The information for nonbasic variables is given in the "Summary of Results" and the information for basic variables is given under the heading "RNGOBJ."
4. Sensitivity analysis for the right-hand sides. The information for nonbinding constraints is given in the "Summary of Results" and for binding constraints under the heading "RNGRHS."
5. The initial tableau is given to help identify the variables that have been added.
6. The final tableau gives the inverse of the basis set of vectors at optimality and also the pertinent information for alterations in the solution due to changes in the right-hand side. It is also necessary for adding or removing constraints.

8-4 SOLUTION RESULTS—ACTUAL SOLUTION

For this specific problem, the actual solution is given in Exhibit 8-2 in the "Summary of Results." Notations and markings have been added to the output to assist in the interpretation, even though the output already lends itself well to interpretation.

Exhibit 8-2

SUMMARY OF RESULTS

VAR NO	VAR NAME	ROW NO	STATUS	ACTIVITY LEVEL	OPPORTUNITY COST	BOUND VALUE	
1	X1	--	B	12.8000000	--	INF	
2	X2	--	B	26.5600000	--	INF	
3	X3	--	B	11.5200000	--	INF	
4	X4	--	B	12.8000000	--	INF	
5	X5	--	B	4.4666667	--	INF	
6	X6	--	B	5.8977778	--	INF	
7	X7	--	B	12.8000000	--	INF	
8	X8	--	B	6.5600000	--	INF	
9	X9	--	B	26.5600000	--	INF	
10	X10	--	B	26.5600000	--	INF	
11	X11	--	B	16.1600000	--	INF	
12	X12	--	B	11.5200000	--	INF	
13	X13	--	B	11.5200000	--	INF	
14	X14	--	B	11.5200000	--	INF	
15	X15	--	NB	--	0.0000000	INF	
16	X16	--	NB	--	0.0000000	INF	
17	X17	--	B	8.3333333	--	INF	
18	X18	--	B	6.9022222	--	INF	
19	X19	--	NB	--	0.0000000	INF	
20	X20	--	B	20.0000000	--	INF	
21	X21	--	NB	--	0.0000000	INF	
22	X22	--	NB	--	0.0000000	INF	
23	X23	--	B	10.4000000	--	INF	
24	X24	--	NB	--	0.0000000	INF	
25	X25	--	NB	--	0.0000000	INF	
26	X26	--	NB	--	0.0000000	INF	
27	X27	--	B	11.5200000	--	INF	
28	--SLACK	4	B	2.8000000	--	INF	
29	--SLACK	5	B	6.5600000	--	INF	
30	--SLACK	6	B	1.5200000	--	INF	
31	--SLACK D-	1	B	17.2000000	--	INF	Sales constraints
32	--SLACK D-	2	B	93.4400000	--	INF	Sales constraints
33	--SLACK D-	3	B	138.4800000	--		Sales constraints
34	--ARTIF D-	4	NB	--	0.0000000		Sales constraints
35	--ARTIF D-	5	NB	--	0.0000000	INF	Sales constraints
36	--ARTIF D-	6	NB	--	0.0000000	INF	Sales constraints
37	--ARTIF D-	7	NB	--	16.3280000	INF	Stage-balance constraints
38	--ARTIF D-	8	NB	--	630.4000000	INF	Stage-balance constraints
39	--ARTIF D-	9	NB	--	90.0000000	INF	Stage-balance constraints
40	--ARTIF D-	10	NB	--	13.2720000	INF	Stage-balance constraints
41	--ARTIF D-	11	NB	--	27.2133333		Stage-balance constraints
42	--ARTIF D-	12	NB	--	73.5466667		Stage-balance constraints
43	--ARTIF D-	13	NB	--	5.0000000		Stage-balance constraints
44	--ARTIF D-	14	NB	--	44.2400000		Stage-balance constraints
45	--ARTIF D-	15	NB	--	13.6066667	INF	Stage-balance constraints
46	--ARTIF D-	16	NB	--	42.0266667	INF	Stage-balance constraints
47	--ARTIF D-	17	NB	--	7.5000000	INF	Stage-balance constraints
48	--ARTIF D-	18	NB	--	36.8666667	INF	Stage-balance constraints
49	--SLACK D-	19	NB	--	.5803333		Day-capacity constraints
50	--SLACK D-	20	NB	--	.5506667		Day-capacity constraints
51	--SLACK D-	21	NB	--	.2000000		Day-capacity constraints
52	--SLACK D-	22	NB	--	.5746667		Day-capacity constraints
53	--SLACK D-	23	NB	--	.3803333		Night-capacity constraints
54	--SLACK D-	24	NB	--	.4506667		Night-capacity constraints
55	--SLACK D-	25	B	257.6000000	--		Night-capacity constraints
56	--SLACK D-	26	NB	--	.5746667		Night-capacity constraints

MAXIMUM VALUE OF THE OBJECTIVE FUNCTION = 6764.800000

CALCULATION TIME WAS .1780 SECONDS FOR 22 ITERATIONS.

There are 56 variables in total for this problem, and these are numbered consecutively in the left-hand column under the heading "VAR NO." In the next column, labeled "VAR NAME," the first 27 entries are x_1 to x_{27}, the original variables in the statement of the problem. In the same column, the slacks and artificials are appropriately identified, surpluses being labeled as slacks.

In the column labeled "ROW NO," the constraints are identified by number, the numbers having been supplied on the input as shown in Exhibit 8-1. A horizontal line has been drawn between variables 30 and 31. By reference to the "ROW NO" column, it can be seen that the line just drawn separates the numbers into two categories: those numbered consecutively from 1 to 26 below that line, and the other numbers, which in this case happen to be 4, 5, and 6. The numbers above the line, in this case 4, 5, and 6, identify the constraints that have surplus variables. The consecutive numbers below that line identify the constraints for easy reference to the dual solutions, which appear in the "OPPORTUNITY COST" column on the same line.

The actual solutions are given in the "ACTIVITY LEVEL" column. Since there are 26 constraints in the problem, there are exactly 26 entries in the column. These are the basic solutions. All dashes in this column refer to nonbasic solutions, which are, of course, zero. We can thus see that the solution here calls for the production of 12.8 *A*'s, 26.56 *B*'s, and 11.52 *C*'s, as shown by the top three entries. The remainder of the entries down through x_{27} give the allocations by stage, by product, by shift. The other entries are slacks or surpluses. Even though dashes represent zeros, it is also possible to have a *printed zero*, which would then be a degenerate solution. To have a real-world solution it will never be possible to have an entry corresponding to an artificial variable, unless it is a printed zero. This means that an artificial variable cannot be a basic variable unless it is at the *zero* level. If the program cannot eliminate the artificial variables at the nonzero level, a statement is printed indicating that the artificial variables cannot be suppressed. Such a printout would indicate that the original statement of the problem needs to be reviewed, for it does not actually possess a solution.

The information in the "OPPORTUNITY COST" cost column will be explained next.

8-5 SOLUTION RESULTS—OPPORTUNITY COSTS

This column lists a value for all nonbasic variables. Consequently, the only entries here are whenever there is *not* an entry in the activity level column. Thus corresponding to each line in the table, either the activity level or the opportunity cost is zero, meaning that the product of the two entries is zero. We sometimes refer to this property as the *complementary slackness* property.

Consider the entries below the horizontal line drawn in Exhibit 8-2. A nonzero entry in this portion indicates that the constraint is binding. This entry has been computed from $\mathbf{c_B B^{-1}}$ and as such not only tells us the marginal rate of change of the objective function but also tells us the direction of movement of the right-hand side to make that gain. For this particular problem there are no negative values, and hence all gains to be made are made by raising the right-hand side.

We can easily classify the constraints for this problem and they are so marked on Exhibit 8-2. The first six are quantity constraints, all of which have zero opportunity costs—the constraints are nonbinding. This could have also been seen from the slacks in the variables 31, 32, and 33 and the surpluses in variables 28, 29, and 30, but the direct method in looking at the opportunity costs is easier and needs no identification as to which variables are slacks or surpluses.

The opportunity costs corresponding to constraints 7 through 18 are all positive but are meaningless in this case. If the right-hand side of any one of these constraints were changed, we would no longer have balanced stage production, so they cannot be changed. These constraints are really definitions and were put in as a convenience in the solution of the problem; they were never candidates for sensitivity analysis.

The constraints on stage capacities are all binding except for the one on stage three at night. The range over which these values hold will be discussed later.

The second classification of opportunity costs are those shown above the horizontal line drawn in Exhibit 8-2. Each of these opportunity costs indicates the penalty that would be incurred if 1 unit of that variable were entered into the program; they are, of course, the $(z_j - c_j)$'s. For this problem there are eight printed entries. It so happens that for this case all the printed entries are *zeros*.

8-6 ALTERNATE OPTIMAL SOLUTIONS

Whenever a zero appears in the opportunity cost column, it indicates that this variable can be entered into the solution at no loss and is therefore an alternate optimal solution. The exception to this is in the case in which the zero corresponds to an artificial variable—which, of course, is not a candidate for entry into the solution.

For our present example there are eight candidates for alternate optimal solutions. Since there are 26 basic variables, these can be selected from among a total of 34 variables, the 26 already in solution and the 8 with zero opportunity costs. The number of combinations of 34 variables taken 26 at a time is in excess of 18,000,000. This indicates that there are many alternate optimal solutions, even though all of them may not be usable. There is no doubt that there are many practical ones among them. Even though each of these solutions would give the same value of the objective function, there may be intangible differences, which would dictate that a selection of the best actual solution is not a simple matter with so many from which to choose. The features of this solution (many alternate solutions) should be kept in mind, because reference to it will be made in Chap. 9.

8-7 SENSITIVITY ON THE RIGHT-HAND SIDE

One of the instructions listed toward the end of Exhibit 8-1 is RNGRHS. This calls for a sensitivity analysis of the right-hand sides of the original 26 constraints. From the "Summary of Results" in Exhibit 8-2 it can be seen that constraints 7 through 24 and 26 are binding, since each of these has a nonzero opportunity cost.

For example, constraint 19, which limits the day stage 1 production capacity to 800 hours, is a binding constraint. Its dual solution of $0.5803 (opportunity cost) indicates that the marginal return to profit for an additional hour of capacity is $0.58033. To find the number of hours for which this rate holds, refer to Exhibit 8-3, which contains the RNGRHS output. Corresponding to constraint 19, labeled under column BI, it is seen that the marginal rate holds up to 819.69 hours (maximum BI) for a new profit of $6776.20, listed right underneath it. The minimum BI is also given, to indicate what happens if the capacity should be reduced to its lower limit. Similar results are given for all binding constraints.

8-8 SENSITIVITY ON COEFFICIENTS OF OBJECTIVE FUNCTION

Sensitivity analysis can also be made on the coefficients of the objective function by referring to Exhibit 8-4. The analysis is made for all coefficients of the objective function for basic variables. For example, if one wishes to examine the effect of varying the selling price of *A* from its value of $750 per 1000, reference is made to the CJ of 1 (coefficient of x_1). The results show that the selling price can vary from a price of $496.50 up to a price of $905.59 without changing the solution. The value of the objective function will be changed, of course, since income would be changed. At the upper limit of $905.59, the increase in income is $905.59 less $750, which is $155.59 per 1000 *A*'s. Since the solution calls for the manufacture of 12,800 *A*'s, the increase in income is $155.59 multiplied by 12.8, to give an increase of approximately $1991.55, yielding a new profit of $8756.40, as indicated. Similar analyses can be made for each of the other basic variables.

Exhibit 8-3

RNGRHS
******
(OPTIMALITY RANGE FOR RIGHT-HAND-SIDE CONSTANTS)
NON-SLACK RESOURCES ONLY

BI	XOUT	MIN BI Z-LOWER	ORIGINAL BI Z	MAX BI Z-UPPER	XOUT
7	8	-4.2008 6696.2	0. 6764.8	.82038 6778.2	30
8	30	-1.5833 5766.7	0. 6764.8	1.3197 7596.7	55
9	55	-1.4311 6636.0	0. 6764.8	6.9022 7386.0	18
10	30	-1.7025 6742.2	0. 6764.8	12.698 6933.3	7
11	29	-2.5205 6696.2	0. 6764.8	.49223 6778.2	30
12	30	-13.571 5766.7	0. 6764.8	11.311 7596.7	55
13	55	-25.760 6636.0	0. 6764.8	26.560 6897.6	10
14	30	-.51075 6742.2	0. 6764.8	4.1935 6950.3	23
15	8	-5.0410 6696.2	0. 6764.8	.98446 6778.2	30
16	30	-23.750 5766.7	0. 6764.8	12.308 7282.1	13
17	55	-17.173 6636.0	0. 6764.8	11.520 6851.2	14
18	30	-.61290 6742.2	0. 6764.8	5.3478 6962.0	8
19	8	699.18 6706.3	800.00 6764.8	819.69 6776.2	30
20	30	4050.0 6241.7	5000.0 6764.8	5791.8 7200.8	55
21	55	1242.4 6713.3	1500.0 6764.8	2742.4 7013.3	18
22	30	584.68 6756.0	600.00 6764.8	725.81 6837.1	23
23	29	699.18 6726.5	800.00 6764.8	819.69 6772.3	30
24	30	4050.0 6336.7	5000.0 6764.8	5791.8 7121.6	55
26	30	584.68 6756.0	600.00 6764.8	733.70 6841.6	8

Exhibit 8-4

RNGOBJ
******
(OPTIMALITY RANGE FOR COST COEFFICIENTS)
BASIC VARIABLES ONLY

CJ	XIN	MIN CJ ------- Z-LOWER	ORIGINAL CJ ----------- Z	MAX CJ ------- Z-UPPER	XIN
18	22	-90.000 6764.8	-90.000 6764.8	-54.000 7013.3	51
23	19	-27.000 6764.8	-27.000 6764.8	-27.000 6764.8	15
27	15	-22.500 6764.8	-22.500 6764.8	-17.573 6821.6	53
1	54	496.50 3520.0	750.00 6764.8	905.59 8756.4	53
4	16	-2.4000 6764.8	-2.4000 6764.8	153.19 8756.4	53
5	50	-609.75 5381.2	-300.00 6764.8	-300.00 6764.8	21
6	51	-90.000 6552.5	-54.000 6764.8	-54.000 6764.8	22
2	53	144.15 6609.5	150.00 6764.8	161.71 7075.9	52
8	49	-12.919 6706.3	-4.0000 6764.8	-4.0000 6764.8	16
9	21	-35.000 6764.8	-35.000 6764.8	-23.288 7075.9	52
10	22	-3.0000 6764.8	-3.0000 6764.8	8.7120 7075.9	52
3	52	94.207 6698.1	100.00 6764.8	104.93 6821.6	53
12	24	-2.0000 6764.8	-2.0000 6764.8	2.9266 6821.6	53
13	25	-20.000 6764.8	-20.000 6764.8	-15.073 6821.6	53
14	26	-4.5000 6764.8	-4.5000 6764.8	.42660 6821.6	53
11	15	-27.000 6764.8	-27.000 6764.8	-27.000 6764.8	19
7	19	-8.1000 6764.8	-8.1000 6764.8	147.49 8756.4	53
17	21	-360.00 6764.8	-360.00 6764.8	*INF*	
20	16	-12.000 6764.8	-12.000 6764.8	*INF*	

For an analysis of the coefficients of nonbasic variables, reference must be made to either the final tableau or to the opportunity cost for that particular variable in the "Summary of Results." The value indicates the amount by which the coefficient must change in order for that variable to be brought into the solution.

8-9 OPTIMAL SOLUTION—BASIS INVERSE

This particular computer program has been organized so that when slack and artificial variables have been added, the identity matrix appears at the extreme right of the initial tableau. Because of this arrangement, we find that the inverse of the basis set of vectors will appear at the extreme right of the optimal tableau. Since there are 26 constraints, this basis inverse will be a 26 by 26 matrix, which because of its size has not been recorded here. The inverse does have some important uses, however. Most important is its value when seeking the actual solution when a particular right-hand side has been altered. $\mathbf{B}^{-1}$ can be multiplied by the altered **b**-vector to produce the new solutions vector.

8-10 COMPUTER CODES FOR LINEAR PROGRAMMING

The present chapter has been written to illustrate the inherent simplicity in using a ready-made program for solving linear programming problems whose mathematical statement has been made. Although Northwestern University's MPOS program has been used, there are many other programs, with perhaps only slightly different characteristics, that can be used just as easily and just as efficiently.

Regardless of which computer code might be used, to use it intelligently, one must have a thorough knowledge of linear programming. The program above has been used for illustrative purposes. All good programs will have essentially the same features, so one must study the format and the features of the program to be used. One should be able to learn the features of a program within a few hours.

The exercises that follow have been constructed so as to further enhance one's use and understanding of a linear programming code.

EXERCISES

The exercises in this chapter are much more comprehensive than those in previous chapters. Each of these will require a computer for its solution. The analyses are intended to be somewhat detailed and to cover a wide range of ideas.

1 The Generation Machine Company directs its sales expense into various endeavors in order to produce sales. One type of activity that is effective is a regional sales conference. There are six regions (designated I to VI) in which these conferences take place weekly, and some monthly. Each of the conferences is for either a full day or a half-day. There is a per person cost for each of these conferences and an expected resulting sales. The conferences available, their costs, and the resulting sales are listed below.

Sales conferences	Cost	Resulting sales
Monthly I —full day	\$120	\$13,500
Weekly II —full day	130	21,500
Weekly II —half-day	80	11,500
Weekly III—full day	60	7,500
Monthly IV—full day	100	11,800
Monthly IV—half-day	60	9,500
Weekly V —full day	200	22,000
Weekly VI—full day	600	97,000
Weekly VI—half-day	350	50,000

In addition to these activities, a TV commercial costing \$1250 would produce \$118,500 in sales.

A newspaper ad in a group of local newspapers would cost \$330 and produce \$57,000 in sales.

Finally, having an inquiry office open for questions for 1 day would cost \$180 and produce \$23,800 in sales.

The company would like to maximize its sales from its sales expense but does wish to impose a few constraints. The plan was to cover a 6-month period, and for that period the sales expense budget is \$52,500. It was further decided that at least half of the budget must go to the weekly and monthly sales conferences. At least one person must be sent to weekly VI and monthly I, each for a full day. The TV commercial would be used at least once, and furthermore, no more than one person would be sent to each of the conferences monthly IV and weekly II.

A number of questions will be asked regarding the solution to this model, but so that each reader interprets the model in the same way, let us consider the model.

Let

x_1 = number of persons sent to monthly I —full day
x_2 = number of persons sent to weekly II —full day
x_3 = number of persons sent to weekly II —half-day
x_4 = number of persons sent to weekly III—full day
x_5 = number of persons sent to monthly IV—full day
x_6 = number of persons sent to monthly IV—half-day
x_7 = number of persons sent to weekly V —full day
x_8 = number of persons sent to weekly VI—full day
x_9 = number of persons sent to weekly VI—half-day
x_{10} = number of TV commercials used
x_{11} = number of newspaper ads
x_{12} = days inquiry office is kept open

The model, then, becomes:

Find

$$x_1, \ldots, x_{12} \geqq 0$$

such that:

Total budget:

$$120x_1 + 130x_2 + 80x_3 + 60x_4 + 100x_5 + 60x_6 + 200x_7 + 600x_8 + 350x_9 + 1250x_{10} + 330x_{11} + 180x_{12} = 52{,}500$$

Half of budget to sales conferences:

$$120x_1 + 130x_2 + 80x_3 + 60x_4 + 100x_5 + 60x_6 + 200x_7 + 600x_8 + 350x_9 \geqq 26{,}250$$

Minimum of weekly VI and monthly I (both full day):

$$x_8 \geqq 1$$

$$x_1 \geqq 1$$

TV commercial requirement:

$$x_{10} \geqq 1$$

No more than one person to each of monthly IV and weekly II (full day):

$$x_5 \leqq 6$$

$$x_2 \leqq 26$$

such that

$$135x_1 + 215x_2 + 115x_3 + 75x_4 + 118x_5 + 95x_6 + 220x_7 + 970x_8 + 500x_9 + 1185x_{10} + 570x_{11} + 238x_{12} \text{ is a } maximum$$

Another person might formulate this model in another way, but questions probably arise only with regard to the last two constraints. The constraint $x_5 \leqq 6$ does not really limit each conference to one person, but it does limit the total to 6 for a 6-month period, which is a liberal interpretation. The final constraint applies in a way similar to weekly II.

Using this model, answer the following questions. Run the model on a computer.

(*a*) Explain the optimal allocation of sales expense. Give some explanation of fractional results.

(*b*) Which constraints are binding? Explain.

(*c*) Compute the sales dollars return per dollar expended for the optimal 6-month plan.

(*d*) The constraint on total budget is written with an equal-to constraint. Explain the meaning of the opportunity cost associated with this constraint. What would the optimal total sales be if the budget were raised to $62,500? Where would the additional $10,000 be allocated? What is the per unit return on the extra $10,000, and why is this different from the figure obtained in part (*c*)?

(*e*) Explain how the requirement that no more than one person to monthly IV and weekly II affects the solution.

(*f*) You will note that variable x_1 is a basic variable in the optimal solution (i.e., $x_1 = 1$) and yields $13,500 in sales. By how much must the $13,500 increase before there is a change in the basic variables? If the increase is larger than this determined amount, what happens to the solution?

(*g*) You will note that variable x_3 is not in the solution. The sales dollar return for an expense of \$80 is presently \$11,500. How much would this return have to be before it would be brought into the solution?

(*h*) Do you think that the requirement that half of the budget be spent on sales conferences has much effect on the return? Explain and justify your answer. What is the return if you allocate only \$16,250 to sales conferences?

(*i*) What would the sales dollar return be if the requirement of the TV commercial were completely eliminated? Where would the \$1250 be allocated (i.e., what is the new solution)?

(*j*) What would the optimal solution be if three TV commercials were required? Give the complete new solution.

(*k*) If you had the choice of eliminating either constraint 3 or 4, which would you eliminate and why?

(*l*) Suppose that you were given the requirement that monthly I had to be attended each of the 6 months (i.e., $x_1 = 6$). Detail the way in which it affects the solution.

(*m*) Suppose that a new direct mail advertising plan is made available to you. The cost is \$200 and the expected return is \$30,000. Should you try this method; should you allocate some budget to it? If so, why so? If not, how much better should the return be in order for you to try it?

2 The Schwartz Investment Group has been asked to determine the exact nature of a \$1,000,000 portfolio for a prospective customer, Mr. Smith. Since Mr. Smith is familiar with the stock market, he has injected a number of ideas into the makeup of the portfolio and has provided Schwartz with a number of restrictions, even though he does admit that some of the restrictions are sort of "seat of the pants." Nevertheless, he has required that no more than 50 percent can be invested in common stocks, and that at least 35 percent be in bonds.

After some preliminary study, the Schwartz group came up with a list of 10 securities from which they thought the investments should be selected, and Mr. Smith agreed. These securities are listed below with identification, price, and expected return, the latter two items being in dollars.

After a review of the list, Mr. Smith noticed that there were several utilities and insisted that at least 20 percent be in utilities; and since there were three airlines, he thought he ought to limit the investment in airlines to 15 percent of the portfolio.

Find the number of shares of each security that will maximize the expected return. There are five constraints to this problem, the first of which is the amount invested. Let us agree to write that as an equal-to-or-less-than constraint with \$1,000,000 on the right-hand side. Each of the other four constraints involves a limitation on the amount invested in a certain type of security. There are two basic ways in which these constraints can be written, the first having the

Security number	Name	Price	Expected return
1	Local Insurance (Common)	\$120	\$8.00
2	Chicago City Bonds	980	43.00
3	Georgia Power Co. (Common)	60	3.50
4	Middle Telephone (Common)	50	4.00
5	Ill. Gas Co. (Preferred)	95	6.00
6	Feeder Airlines (Common)	70	7.00
7	Shuttle Airlines (Common)	44	4.00
8	Flutter Airlines (Common)	125	10.00
9	California Bonds	1015	45.00
10	Machine Industries (Common)	24	2.00

right-hand sides in actual dollar amounts and the second having the right-hand sides as a percentage times the total invested. We shall call these model 1 and model 2. To illustrate and clarify, consider the form of the common stock constraint for each of the two models.

The total amount invested in the portfolio will be

$$120x_1 + 980x_2 + 60x_3 + 50x_4 + 95x_5 + 70x_6 + 44x_7 + 125x_8 + 1015x_9 + 24x_{10}$$

The total amount invested in utilities is

$$60x_3 + 50x_4 + 95x_5$$

Since we are insisting that at least 20 percent be in utilities, we shall set the right-hand side of this constraint in model 1 to be 20 percent of \$1,000,000, or \$200,000, and the constraint would be

$$60x_3 + 50x_4 + 95x_5 \geqq 200{,}000$$

For model 2 the right-hand side will be 20 percent of the total investment, and we write

$$60x_3 + 50x_4 + 95x_5 \geqq 0.20(120x_1 + 980x_2 + 60x_3 + 50x_4 + 95x_5 + 70x_6 + 44x_7 + 125x_8 + 1015x_9 + 24x_{10})$$

which reduces to

$$-24x_1 - 196x_2 + 48x_3 + 40x_4 + 76x_5 - 14x_6 - 8.8x_7 - 25x_8 - 203x_9 - 4.8x_{10} \geqq 0$$

Each of the other constraints would be handled in a similar way. Thus we might refer to model 1 as the model having fixed dollars on the right-hand side and model 2 as having proportionate dollars on the right-hand side.

Solve each of these models on the computer. The actual solution will be the same, of course, but there will be some differences in the sensitivity analysis. Using model 1, consider the following:

(*a*) Give the solution, allowing fractional shares, and note that with five constraints there will be only five securities selected.

(*b*) Which of the constraints are binding?

(*c*) What is the rate of return for the entire portfolio? What is the rate of return for each additional dollar added to the portfolio, keeping the remainder of the constraints as is? For how many dollars added to the portfolio, will this latter rate of return hold?

(*d*) What kind of gains could be made by allowing more to be invested in airlines? What would the portfolio dollar return be if you were allowed to raise the allowable percentage to 20 percent?

(*e*) How much would be gained by requiring only 30 percent to be in bonds?

(*f*) What is the rate of return for an additional dollar added to the portfolio, and why is this rate not the same as the overall return of 6.899 percent? Why is it lower than the overall return?

(*g*) Variable x_4, Telephone Company, is in the solution. It has a return of \$4 per share. What is the range over which this return can vary and still keep the same mix in the portfolio? Suppose that the return is \$3.50 instead of \$4. What will the return be on the complete portfolio?

(*h*) By how much could the return on Machine Industries drop from the present \$2 before it drops out of the portfolio?

(*i*) What would the return on Local Insurance have to be before it would be brought into the solution?

(*j*) Referring to model 2, investigate the rate of return for an additional dollar added to the portfolio. Is this the same as for model 1? If not, why not? If so, why so? Is the range the same as for model 1? If so, why so? If not, why not?

(*k*) Constraint 2 refers to common stocks. If the upper limit is $600,000 instead of $500,000, what will the new return be, and what is the exact composition of the portfolio (number of shares)?

3 Consider the following problem concerned with the loading of available cargo into a ship. Three different commodities are available, and these must be loaded into the forward, center, and aft sections of the hold, each of which has a weight capacity as well as a volume capacity. In addition, there are some constraints regarding the balance or trim of the ship. The following information describes the constraints for loading the ship.

Section of the ship	Weight capacity, tons	Volume capacity, ft^3
Forward	2000	50,000
Center	3000	67,500
Aft	1500	15,000

The following cargos are available and the shipowners may accept all or any part of the commodity.

Commodity	Amount, tons	Volume/ton, ft^3	Profit/ton
P	6000	30	$12
Q	4000	25	16
R	2000	12.5	10

To keep the ship in balance, no section of the ship may be over- or underloaded. Specifically, the load in any hold must be proportional to the capacity of the hold. Call this model 1.

For model 2 consider that there is some variation allowed in this trim constraint. Instead of requiring that the center weight be 1.5 times the forward weight, let the requirement be that center weight = 1.5 (forward weight) ± 0.05 (forward weight) and that center weight = 2 (aft weight) ± 0.05 (aft weight).

What cargo should be accepted and where should it be loaded to maximize profits? Find these for both models.

For model 1:

(*a*) Give the solution for the optimal cargo and exactly where it is loaded. Is there anything unusual about this result—alternates, degeneracies, and so on?

(*b*) Which constraints are binding?

(*c*) If you could have more of one of the cargos made available to you, which one would you select? Why?

(*d*) Suppose that you could remodel the ship and make more space available in each of the holds, without altering the basic characteristics of the ship. Which one would you alter to increase the space? Why? How much additional space would you try to get?

(*e*) Find the range of profit for each of the commodities that would maintain the same ship loading. Each is to be done separately.

(*f*) Note that the volume in the aft section is a limitation to profit. What would it be worth to have additional aft space? For what amount of space will this rate hold? (Be careful!)

(*g*) None of product *P* is in the cargo. How much would its profit have to improve before it would be accepted? Where would it go?

(*h*) Interpret the meaning of the dual variables for the trim constraints.

For model 2:

(*i*) How much is the profit improved by relaxing the trim constraints?

(*j*) Are there any alternates to this solution? Give them.

(*k*) Would the solution improve if you could get more of any of the cargos. Which one? How much more would you like to get? What would be the improvement in profit?

(*l*) Indicate how you would further like to relax the trim constraints. What kinds of gains could you make?

$\frac{3}{2}$

Max. $12PF + 12PC + 12PA + 16QF + 16QC + 16QA + 10RF + 10RC + 10RA$

$2(PF + QF + RF) = 3(PC + QC + RC)$

$2(PF + QF + RF) = 1.5(PA + QA + RA)$

$30PF + 25QF + 12.5RF \le 50000$

$30PC + 25QC + 12.5RC \le 67500$

$36PA + 25QA + 12.5RA \le 15000$

$PF + PC + PA \le 6000$

$QF + QC + QA \le 4000$

$RF + RC + RA \le 2000$

≥ 0

CHAPTER

NINE

IMPORTANCE OF MODELING FOR INTERPRETATION OF LINEAR PROGRAMMING PROBLEMS

9-1 INTRODUCTION

It is often thought to be very efficient to solve a linear programming problem by using the fewest number of variables. If one is interested only in the solution and not in any kind of sensitivity analysis, this may be true, but if one is interested in the effect of a variation of the parameters of the problem, many more details may be required to enhance the analysis and interpretation.

9-2 SCHEDULING PROBLEM

In order to consider the benefits of efficient modeling, consider the production scheduling problem shown in Fig. 9-1 (no claim is made for the originality of the example).

A manufacturing plant makes two products, *A* and *B*, which are routed through four processing centers, 1, 2, 3, and 4, as shown by the solid lines in Fig. 9-1. If there is spare capacity in center 3, it is possible to route product *A* through center 3 (dashed line) instead of going through center 2 twice, but this is more expensive.

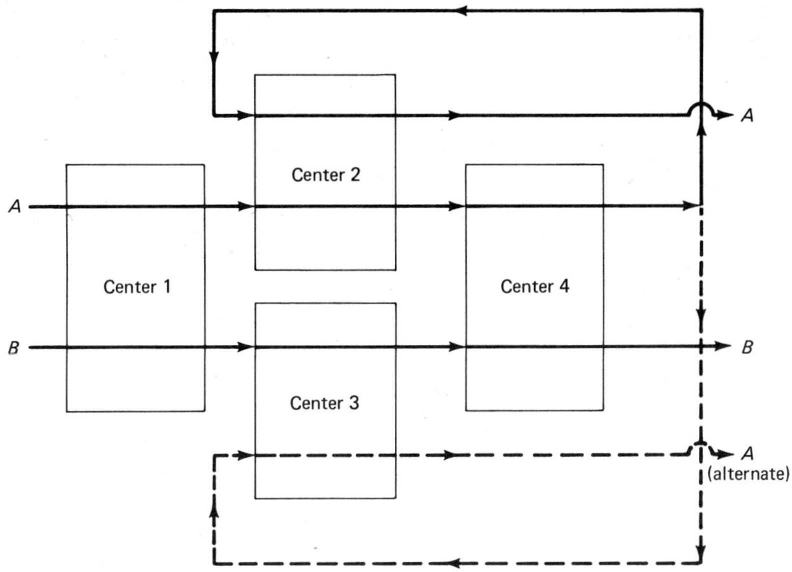

Figure 9-1 Production scheduling.

Given the information in Tables 9-1 and 9-2, how should production be scheduled so as to maximize profit? By production schedule is meant the specification of the following three amounts:

1. The daily amount of raw material used for *A*, regular route.
2. The daily amount of raw material used for *A*, optional route.
3. The daily amount of raw material used for *B*.

Assume that sufficient in-process storage capacity is available at no additional cost and that no two things can be processed through any center at the same time.

Table 9-1

Product	center	Input, gal/h	Recovery, %	Running cost, $/h
A	1	300	90	150
	2 (1st pass)	450	95	200
	4	250	85	180
	2 (2nd pass)	400	80	220
	3	350	75	250
B	1	500	90	300
	3	480	85	250
	4	400	80	240

Table 9-2

Product	Raw-material cost, \$/gal	Sales price, \$/finished gallon	Maximum daily sales, gal finished product
A	5	20	1700
B	6	18	1500

Additional constraints:

1. Centers 1 and 4 run up to 16 h/day.
2. Centers 2 and 3 run up to 12 h/day.
3. Shipping facilities limit the daily output of *A* and *B* together to a total of 2500 gal.

9-3 INITIAL INTUITIVE MODEL FORMULATION

Consider first the formulation of the problem in a form in which as few variables as possible are used. Therefore,

x_1 = number of gallons of input of *A*, which is to follow the regular processing route
x_2 = number of gallons of input of *A*, which is to follow the alternate processing route
x_3 = number of gallons of input of *B*, which has only the one route

In Fig. 9-2 these unknowns are shown with both the inputs to and the outputs from each of the centers. Each of the outputs is expressed in terms of the original inputs x_1, x_2, and x_3, having been modified by the appropriate recovery rates. On each path through a particular center, there appear two numbers; the left-hand number is the number of gallons per hour that can be processed and the right-hand number is the dollar cost per hour for processing.

To further clarify Fig. 9-2, let us trace raw material *A* through the regular route. Starting with an amount x_1 entering center 1, the amount coming out of that center is $0.9x_1$, since there is a 90 percent recovery rate. With $0.9x_1$ entering center 2 and a 95 percent recovery rate, there will be $0.95 \times 0.9x_1 = 0.855x_1$ coming out of center 2; and so on. Since x_1 units enter center 1, it will take $x_1/300$ h for it to be processed and it will cost $\$150(x_1/300)$. Similar use of the other numbers in Fig. 9-2 enable one to arrive at the model developed on p. 185.

Reference to Fig. 9-3 on p. 189 might help to clarify the foregoing production process.

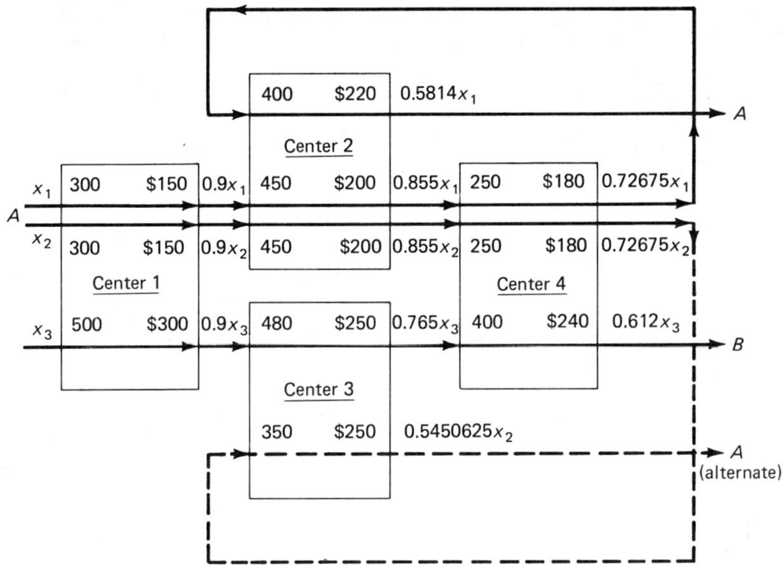

Figure 9-2 Production scheduling with data.

9-4 MATHEMATICAL STATEMENT OF THE PROBLEM

Consider the hourly constraint on center 1. Converting the gallons of input for x_1, x_2, and x_3 into hours, the constraint on center 1 becomes

$$\frac{x_1}{300} + \frac{x_2}{300} + \frac{x_3}{500} \leqq 16$$

When written in this form, the elements are in units of hours, and consequently, any analysis of the right-hand side would be in the units of the effect per hour of change of the right-hand side. To preserve this, the left-hand side would have to be cleared of fractions by writing the constraint in the form

$$0.0033333x_1 + 0.003333x_2 + 0.002x_3 \leqq 16$$

When we do this, we lose something in accuracy and really do not know how many decimals we should carry.

We can also clear the constraints of fractions by multiplying the entire constraint by the least common denominator which is 1500. When we do this, the result is

$$5x_1 + 5x_2 + 3x_3 \leqq 24{,}000$$

While this itself has no rounding error, it does have the disadvantage that a 1-unit change in the right-hand side is now one-fifteen hundredth of an hour instead of 1 hr. We then have to weigh whether we want the accuracy or the convenient units. We have chosen to preserve the accuracy and handle the units problem during the sensitivity analysis. Constraints on the other three centers have been handled in a similar fashion.

The constraints involving sales' limitations and dock capacity are perfectly straightforward. No rounding is necessary.

The profit function involves the compilation of a number of terms. The profit is determined by computing income less cost. The cost is a combination of the raw-material cost and the costs of processing through each of the centers. Consider the terms involving the variable x_1. Since the amount of final product which follows the regular route for A is $0.5814x_1$, the income is 20 times $0.5814x_1$ for a total of $11.628x_1$. The costs are as follows:

$$\text{Raw material} = 5x_1$$

$$\text{Processing through center 1 (1st pass)} = 150\left(\frac{x_1}{300}\right) = 0.5x_1$$

$$\text{Processing through center 2 (1st pass)} = 200\left(\frac{0.9x_1}{450}\right) = 0.4x_1$$

$$\text{Processing through center 4} = 180\left(\frac{0.855x_1}{250}\right) = 0.6156x_1$$

$$\text{Processing through center 2 (2d pass)} = 220\left(\frac{0.72675x_1}{400}\right) = 0.3997125x_1$$

The total of all of the cost factors is $6.9153125x_1$ so that the net profit for A going the regular route is $4.7126875x_1$. A similar computation is made for the coefficients of x_2 and x_3, so that the function to be maximized is

$$4.7126875x_1 + 3.866543x_2 + 3.48825x_3$$

The coefficient of x_2 is not exact but those of x_1 and x_3 have not been rounded. From the above the resulting problem statement is then:

Find

$$x_1, x_2, x_3 \geqq 0$$

such that

$$
\text{Center capacities}
\begin{cases}
5x_1 + 5x_2 + 3x_3 \leqq 24{,}000 \\
13.74075x_1 + 7.2x_2 \leqq 43{,}200 \\
34.884x_2 + 31.5x_3 \leqq 201{,}600 \\
6.84x_1 + 6.84x_2 + 3.825x_3 \leqq 32{,}000
\end{cases}
$$

$$
\text{Sales limits}
\begin{cases}
0.5814x_1 + 0.5450625x_2 \leqq 1700 \\
0.612x_3 \leqq 1500
\end{cases}
$$

$$
\text{Shipping capacity}\{0.5814x_1 + 0.5450625x_2 + 0.612x_3 \leqq 2500
$$

and such that

$$
4.7126875x_1 + 3.866543x_2 + 3.48825x_3 \text{ is a maximum}
$$

9-5 OPTIMAL SOLUTION

Using the mathematical statement above, the computer was used to find the optimal solution. Exhibit 9-1 shows the computer input for this problem, and Exhibit 9-2 gives the summary of results. The solution reveals the following information about the production schedule:

Exhibit 9-1

```
***** PROBLEM NUMBER   1 *****

      *PRODUCTION PROBLEM - SMALL MODEL
      VARIABLES
      X1 TO X3


      MAXIMIZE
      4.7126875X1+3.866543X2+3.48825X3
      CONSTRAINTS
      *CENTER CAPACITIES
 1.   5X1+5X2+3X3.LE.24000
 2.   13.74075X1+7.2X2.LE.43200
 3.   34.884X2+31.5X3.LE.201600
 4.   6.84X1+6.84X2+3.825X3.LE.32000
      *SALES LIMITS
 5.   .5814X1+.5450625X2.LE.1700
 6.   .612X3.LE.1500
      *SHIPPING CAPACITY
 7.   .5814X1+.5450625X2+.612X3.LE.2500
      PRINT
      RNGOBJ
      RNGRHS
      OPTIMIZE
```

Exhibit 9-2

SUMMARY OF RESULTS

VAR NO	VAR NAME	ROW NO	STATUS	ACTIVITY LEVEL	OPPORTUNITY COST	BOUND VALUE
1	X1	--	B	2923.9766082	--	INF
2	X2	--	NB	--	.5516015	INF
3	X3	--	B	1307.1895425	--	INF
4	--SLACK D-	1	B	5458.5483316	--	INF
5	--SLACK D-	2	B	3022.3684211	--	INF
6	--SLACK D-	3	B	160423.5294118	--	INF
7	--SLACK D-	4	B	7000.0000000	--	INF
8	--SLACK D-	5	NB	--	2.4060028	INF
9	--SLACK D-	6	B	700.0000000	--	INF
10	--SLACK D-	7	NB	--	5.6997549	INF

MAXIMUM VALUE OF THE OBJECTIVE FUNCTION = 18339.591933

CALCULATION TIME WAS .0060 SECONDS FOR 2 ITERATIONS.

1. 2923.98 gal follow the regular route for *A*.
2. Nothing follows the alternative route for *A*.
3. 1307.19 gal follow the route for *B*.
4. The net profit is $18,339.59.
5. The actual output for neither *A* nor *B* is revealed directly, although each can be obtained by some additional simple arithmetic.
6. From the dual solutions, one sees that the sales restriction on *A* and the shipping capacity are both limitations on profit. Per unit gains to be made by relaxing either of these constraints is also given by the dual solutions.
7. Nothing in the solution is revealed directly concerning the effect of changing either the rates of gallon throughput for each of the centers or the hourly costs of processing in each of the centers.

9-6 EFFICIENCY OF THE MODEL

This form of the model seems to be relatively neat, and one might even pride oneself on the brevity of the format, thinking restriction of the problem to three variables and seven constraints to be a minor triumph. Some of the shortcomings of the model are suggested in items 5, 6, and 7 in Sec. 9-5. There are others, however.

Exhibit 9-3 gives the sensitivity on both the right-hand side and the objective function. Constraints 5 and 7 are analyzed, but if the model were formulated in another way, perhaps it would reveal something about the effects of changing rates of processing or costs of processing in each of

Exhibit 9-3

RNGRHS
******
(OPTIMALITY RANGE FOR RIGHT-HAND-SIDE CONSTANTS)
NON-SLACK RESOURCES ONLY

BI	XOUT	MIN BI / Z-LOWER	ORIGINAL BI / Z	MAX BI / Z-UPPER	XOUT
5	9	1000.00	1700.0	1827.9	5
		16655.	18340.	18647.	
7	3	1700.0	2500.0	3200.0	9
		13780.	18340.	22329.	

CALCULATION TIME WAS .017 SECONDS

RNGOBJ
******
(OPTIMALITY RANGE FOR COST COEFFICIENTS)
BASIC VARIABLES ONLY

CJ	XIN	MIN CJ / Z-LOWER	ORIGINAL CJ / Z	MAX CJ / Z-UPPER	XIN
1	2	4.1243	4.7127	*INF*	
		16619.	18340.		
3	10	.14211E-13	3.4882	4.9607	8
		13780.	18340.	20264.	

CALCULATION TIME WAS .017 SECONDS

the centers. In ranging the objective function, one can study the effects of changing the coefficients of x_1, x_2, and x_3, but such an analysis is not detailed enough, since the coefficients themselves have been obtained by combining a number of cost factors. For example, 4.7126875, the coefficient of x_1, is an accumulation of dollars of income, costs of raw material, and costs of processing in several of the centers. The effects of varying this figure do not enable you to pinpoint a change or make assignments of such changes to any one of the factors which are combined to make up the final figure.

It would appear that a manager interested in producing the most effective schedule would require a sensitivity analysis that would enable him to study the effects of changes in processing rates or processing costs for each of the centers. With these many factors in mind, a different model for the same problem is developed. This model will enable the manager to learn considerably more about the scheduling operation, and in turn will provide him with information to enable him to be a more effective manager.

9-7 ALTERNATIVE STATEMENT OF THE SCHEDULING PROBLEM

Figure 9-3 is the same flow diagram for the scheduling problem, but with a completely different set of variables. x_1 and x_9 are used to represent the amounts of raw material input for A and B, respectively. x_1 is modified by recovery rates, with a series of new variables indicating the successive outputs from each of the centers along the processing stream. Thus variables x_1 through x_8 refer to product A and variables x_9 through x_{12} refer to product B. Figure 9-3 also shows that a solution for x_7 gives the final output of A going the regular route, and a solution for x_8 gives the final output of A going the alternative route. Similarly x_{12} gives the final output for B, which has only the one route.

Furthermore, the definitions of these variables serve as constraints which involve the recovery rates. The definitions of variables x_{13} through x_{20} serve as constraints which define the hourly processing rates of the centers. The recovery- and hourly-processing-rate definitions and the corresponding constraints are as follows:

Recovery definitions	*When written as constraints*
$x_2 = 0.90x_1$	$x_2 - 0.90x_1 = 0$
$x_3 = 0.95x_2$	$x_3 - 0.95x_2 = 0$
$x_4 = 0.85x_3$	$x_4 - 0.85x_3 = 0$
$x_7 = 0.80x_5$	$x_7 - 0.80x_5 = 0$
$x_8 = 0.75x_6$	$x_8 - 0.75x_6 = 0$
$x_{10} = 0.90x_9$	$x_{10} - 0.90x_9 = 0$
$x_{11} = 0.85x_{10}$	$x_{11} - 0.85x_{10} = 0$
$x_{12} = 0.80x_{11}$	$x_{12} - 0.80x_{11} = 0$

Center processing times	*When written as constraints*
$x_{13} = x_1/300$	$x_1 - 300x_{13} = 0$
$x_{14} = x_2/450$	$x_2 - 450x_{14} = 0$
$x_{15} = x_3/250$	$x_3 - 250x_{15} = 0$
$x_{16} = x_5/400$	$x_5 - 400x_{16} = 0$
$x_{17} = x_6/350$	$x_6 - 350x_{17} = 0$
$x_{18} = x_9/500$	$x_9 - 500x_{18} = 0$
$x_{19} = x_{10}/480$	$x_{10} - 480x_{19} = 0$
$x_{20} = x_{11}/400$	$x_{11} - 400x_{20} = 0$

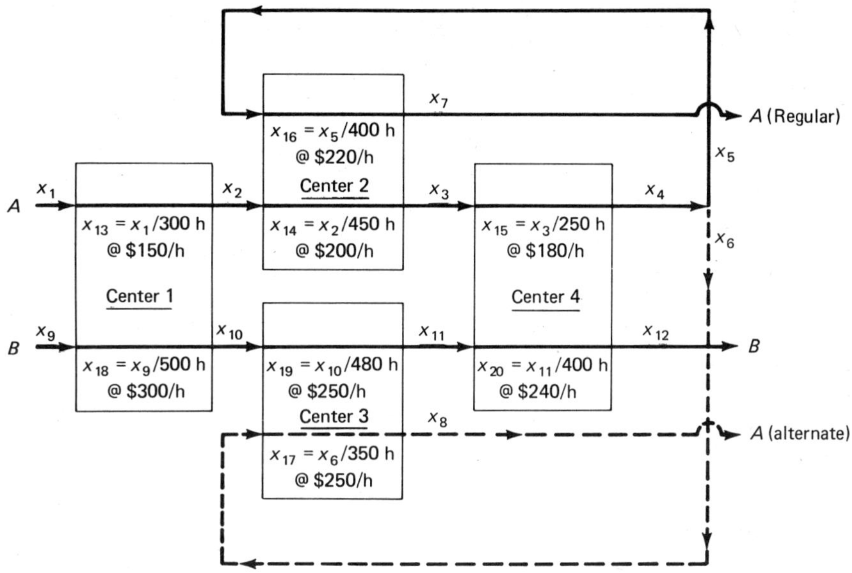

Figure 9-3 Production scheduling—Alternate presentation.

We add the constraint which expresses the balance between A regular and A alternative route as

$$x_4 = x_5 + x_6$$

If we then add the seven constraints of our smaller model in terms of the variables above, we then have a model with 24 constraints and 20 variables. The problem is then stated as follows:

Find

$$x_1 \text{ to } x_{20} \geqq 0$$

such that

$$\text{Recovery definitions} \left\{ \begin{aligned} x_2 &- 0.90x_1 = 0 \\ x_3 &- 0.95x_2 = 0 \\ x_4 &- 0.85x_3 = 0 \\ x_7 &- 0.80x_5 = 0 \\ x_8 &- 0.75x_6 = 0 \\ x_{10} &- 0.90x_9 = 0 \\ x_{11} &- 0.85x_{10} = 0 \\ x_{12} &- 0.80x_{11} = 0 \end{aligned} \right.$$

$$\text{Hourly processing rates}\begin{cases} x_1 - 300x_{13} = 0 \\ x_2 - 450x_{14} = 0 \\ x_3 - 250x_{15} = 0 \\ x_5 - 400x_{16} = 0 \\ x_6 - 350x_{17} = 0 \\ x_9 - 500x_{18} = 0 \\ x_{10} - 480x_{19} = 0 \\ x_{11} - 400x_{20} = 0 \end{cases}$$

$$\text{Balance}\{x_4 - x_5 - x_6 = 0$$

$$\text{Center capacities}\begin{cases} x_{13} + x_{18} \leqq 16 \\ x_{14} + x_{16} \leqq 12 \\ x_{17} + x_{19} \leqq 12 \\ x_{15} + x_{20} \leqq 16 \end{cases}$$

$$\text{Sales constraints}\begin{cases} x_7 + x_8 \leqq 1700 \\ x_{12} \leqq 1500 \end{cases}$$

$$\text{Shipping capacity}\{x_7 + x_8 + x_{12} \leqq 2500$$

and such that

$$\begin{aligned} &-5x_1 + 20x_7 + 20x_8 - 6x_9 + 18x_{12} - 150x_{13} \\ &-200x_{14} - 180x_{15} - 220x_{16} - 250x_{17} - 300x_{18} \\ &-250x_{19} - 240x_{20} \text{ is a maximum} \end{aligned}$$

9-8 SCHEDULING PARAMETERS

The actual parameters for this problem in either form are:

1. Sales limits on *A* and *B*.
2. Shipping capacity.
3. Hourly capacities for centers 1, 2, 3, and 4.
4. Cost of raw material input *A* and *B*.
5. Income from sales of *A* and *B*.
6. Center 1

   Rate of processing (gal/h) } for *A*
   Cost of processing ($/h) }
   Same as above for *B*

7. Center 2

   Rate of processing (gal/h) } for *A*
   Cost of processing ($/h) } first pass
   Same as above for *A* second pass

8. Center 3

   Rate of processing (gal/h) } for *B*
   Cost of processing ($/h) }
   Same as above for *A* alternative route

9. Center 4

   Rate of processing (gal/h) } for *A*
   Cost of processing ($/h) }
   Same as above for *B*

9-9 ADVANTAGES OF THE SECOND MODEL

When using the initial short form of problem, we are able to make a sensitivity study for items 1, 2, and 3 only as outlined in Sec. 9.8. In the larger or expanded form of the model, we are able to analyze all items on the list. The two models for this production problem illustrate very clearly that management must participate in the formulation of a model in order to get the most out of the results. It really is not satisfactory to have management just hand the problem to a project team and ask for results.

Rarely do users just want a solution; they also want to be able to answer a number of "what if" questions. Because of this, any problem solver should approach the solution from the point of view of being fortified to answer such questions. Therefore, in preparing solutions, users should keep in mind that they would like to be able to answer as many questions as possible about the variation of the parameters in the problem. Think of it as entering a management meeting prepared to answer a large percentage of the questions that will be asked of you without having to go back to formulate another model and get another computer solution. If this can be done, you have a good model. (See Exhibit 9-4 for computer input for the large model.)

9-10 SENSITIVITY ON RECOVERY RATES

To see the advantages of the second model, consider the effect of changing the recovery rate for material *A* going through center 1. The recovery rate

Exhibit 9-4

```
***** PROBLEM NUMBER   1 *****

          *PRODUCTION PROBLEM - LARGE MODEL
          VARIABLES
          X1 TO X20
          MAXIMIZE
          -5X1+20X7+20X8-6X9+18X12-150X13-200X14-180X15
          -220X16-250X17-300X18-250X19-240X20
          CONSTRAINTS
          *RECOVERY DEFINITIONS
     1.   X2-.90X1=0
     2.   X3-.95X2=0
     3.   X4-.85X3=0
     4.   X7-.80X5=0
     5.   X8-.75X6=0
     6.   X10-.90X9=0
     7.   X11-.85X10=0
     8.   X12-.80X11=0
          *HOURLY PROCESSING RATES
     9.   X1-300X13=0
    10.   X2-450X14=0
    11.   X3-250X15=0
    12.   X5-400X16=0
    13.   X6-350X17=0
    14.   X9-500X18=0
    15.   X10-480X19=0
    16.   X11-400X20=0
          *BALANCE CONSTRAINT
    17.   X4-X5-X6=0
          *CENTER CAPACITIES
    18.   X13+X18.LE.16
    19.   X14+X16.LE.12
    20.   X17+X19.LE.12
    21.   X15+X20.LE.16
          *SALES CONSTRAINTS
    22.   X7+X8.LE.1700
    23.   X12.LE.1500
          *SHIPPING CAPACITY
    24.   X7+X8+X12.LE.2500
          PRINT
          RNGOBJ
          RNGRHS
          OPTIMIZE
```

for this process is initially 90 percent. It is rather obvious that increasing the recovery rate will improve the profit if nothing else is changed, but what are the effects quantitatively?

Consider the definition equation,

$$x_2 = 0.90x_1$$

and its corresponding constraint,

$$x_2 - 0.90x_1 = 0$$

The dual solution corresponding to this constraint is (Exhibit 9-5)

$$+6.11111111$$

This is interpreted to mean that if the constraint were written

$$x_2 - 0.90x_1 = b_1$$

the instantaneous rate of change of profit is 6.11111 for an unit increase in b_1.

Since the recovery rate is the coefficient of x_1, it is desirable to relate the change in recovery rate to a change in b_1, which is directly related to the dual variable. It can be seen that increasing b_1 makes it possible to get the same output x_2 with less input x_1 (i.e., an increase in recovery rate); b_1 is measured in gallons and the rate of increase in profit is in dollars per gallon.

Using the standard Δ (delta) notation to indicate change,

$$\Delta(x_2 - 0.90x_1) = \Delta b_1$$

$$\Delta x_2 - \Delta\ (0.90x_1) = \Delta b_1$$

$$\Delta x_2 - 0.90\ (\Delta x_1) = \Delta b_1$$

Exhibit 9-5

SUMMARY OF RESULTS

VAR NO	VAR NAME	ROW NO	STATUS	ACTIVITY LEVEL	OPPORTUNITY COST	BOUND VALUE
1	X1	--	B	2923.9766082	--	INF
2	X2	--	B	2631.5789474	--	INF
3	X3	--	B	2500.0000000	--	INF
4	X4	--	B	2125.0000000	--	INF
5	X5	--	B	2125.0000000	--	INF
6	X6	--	B	0.0000000	--	INF
7	X7	--	B	1700.0000000	--	INF
8	X8	--	B	0.0000000	--	INF
9	X9	--	B	1307.1895425	--	INF
10	X10	--	B	1176.4705882	--	INF
11	X11	--	B	1000.0000000	--	INF
12	X12	--	B	800.0000000	--	INF
13	X13	--	B	9.7465887	--	INF
14	X14	--	B	5.8479532	--	INF
15	X15	--	B	10.0000000	--	INF
16	X16	--	B	5.3125000	--	INF
17	X17	--	NB	--	265.6492411	INF
18	X18	--	B	2.6143791	--	INF
19	X19	--	B	2.4509804	--	INF
20	X20	--	B	2.5000000	--	INF
21	--ARTIF D-	1	NB	--	6.1111111	INF
22	--ARTIF D-	2	NB	--	6.9005848	INF
23	--ARTIF D-	3	NB	--	8.9653939	INF
24	--ARTIF D-	4	NB	--	11.8942423	INF
25	--ARTIF D-	5	NB	--	11.8942423	INF
26	--ARTIF D-	6	NB	--	7.3333333	INF
27	--ARTIF D-	7	NB	--	9.2401961	INF
28	--ARTIF D-	8	NB	--	12.3002451	INF
29	--ARTIF D-	9	NB	--	.5000000	INF
30	--ARTIF D-	10	NB	--	.4444444	INF
31	--ARTIF D-	11	NB	--	.7200000	INF
32	--ARTIF D-	12	NB	--	.5500000	INF
33	--ARTIF D-	13	NB	--	-.0447121	INF
34	--ARTIF D-	14	NB	--	.6000000	INF
35	--ARTIF D-	15	NB	--	.5208333	INF
36	--ARTIF D-	16	NB	--	.6000000	INF
37	--ARTIF D-	17	NB	--	-8.9653939	INF
38	--SLACK D-	18	B	3.6390322	--	INF
39	--SLACK D-	19	B	.8395468	--	INF
40	--SLACK D-	20	B	9.5490196	--	INF
41	--SLACK D-	21	B	3.5000000	--	INF
42	--SLACK D-	22	NB	--	2.4060028	INF
43	--SLACK D-	23	B	700.0000000	--	INF
44	--SLACK D-	24	NB	--	5.6997549	INF

MAXIMUM VALUE OF THE OBJECTIVE FUNCTION = 18339.591933

CALCULATION TIME WAS .1160 SECONDS FOR 19 ITERATIONS.

and since x_2 is being held constant, it is true that

$$-0.90(\Delta x_1) = \Delta b_1$$

and
$$\Delta x_1 = \frac{-\Delta b_1}{0.90} = -1.1111(\Delta b_1)$$

Thus lowering x_1 (indicated by the negative sign) by an amount 1.1111 times the increase of b_1 has the same effect on profits as the increase in b_1. This is logical since reducing x_1, the input, results in a savings of \$5 per unit in raw material and a savings of \$0.50 per unit in processing costs. The total per unit savings is then

$$1.1111(5 + 0.5) = 1.1111(5.5) = 6.1111$$

which checks with the value of the dual variable.

To translate this to a change of 1 percent in recovery rate, an approximation for the effects can be made by changing b_1 by an amount equal to 1 percent of x_1, since

$$x_2 - 0.90x_1 = b_1$$

with
$$b_1 = 0.01x_1$$

yields the same value for x_2 as

$$x_2 = 0.91x_1$$

for any given value of x_1. The change in profit for these two cases is not exactly the same, however. The optimal solution to our problem as initially stated yields

$$x_1 = 2924.0$$

$$x_2 = 2631.6$$

Thus, whether we make a change of $0.01x_1$ in b_1 or a change in the recovery rate from 90 to 91 percent, we produce 2631.6 as the output of center 1. A reduction of 1 percent in input of 29.240 substituted for b_1 yields

$$2631.6 - 0.9x_1 - 29.240 = 0$$

or
$$x_1 = 2891.51$$

processed at 90 percent recovery rate.

If we use the constraint

$$x_2 - 0.91x_1 = 0$$

then
$$2631.6 - 0.91x_1 = 0$$

yields

$$x_1 = 2891.87$$

a difference of 0.36 gal at \$6.11 for a total difference of \$2.19. Thus if we use changes in b_1 in the amount of 1 percent of x_1 and use the dual value for the constraint, the expected improvement in profit is approximated by

$$(0.01)(2924)(6.1111) = \$178.69$$

The new profit would then be (approximately)

$$\$18{,}339.59 + \$178.69 = \$18{,}518.28$$

A solution of the problem at the 91 percent recovery rate yields a profit of \$18,516.32. The error is small but, most important, the changes in profit are in the right direction and the managerial implications of a change in recovery rate are of the correct order of magnitude.

The difference between the two cases is illustrated below. The approximation used provides for an input of 2891.51 gal, processed at 90 percent recovery rate to yield 2602.36 gal, to which 29.24 (1 percent of x_1) gallons is added to obtain 2631.60 gal for the input to the next center.

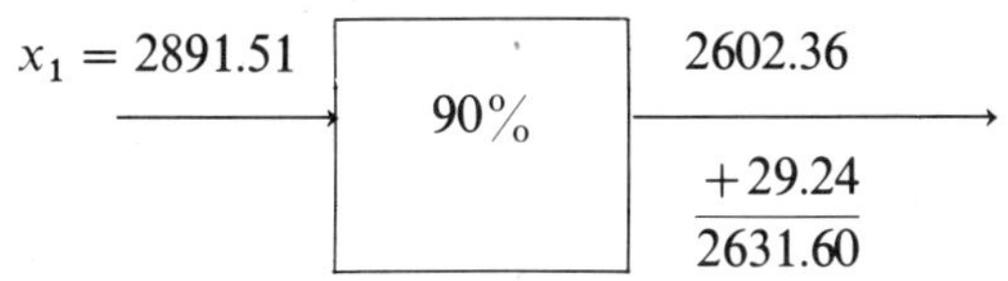

The exact situation would provide for an input of 2891.87 gal, processed at 91 percent recovery rate to yield the 2631.60 gal directly for entry into the next center.

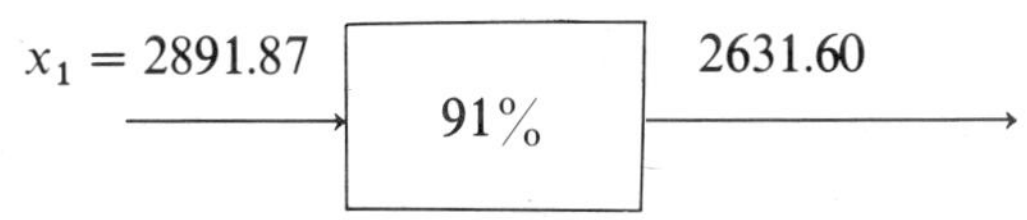

As indicated previously, the difference in the exact situation and the approximation is due to the savings on raw-material cost and processing cost of the difference of 0.36 gal. This amounts to \$2.19.

In a similar fashion, all other recovery-rate changes can be assessed by using their corresponding dual solutions. It is only necessary to change the input to any center by 1 percent and multiply this change by its corresponding dual solution to obtain an estimate of the effect of a change of 1 percent in the value of any recovery rate.

If one wished to find a more exact effect of the change of recovery rate, it must be done by finding the effect of changing the proper coefficient in a related constraint. For the example used here, this amounts to changing the coefficient of the variable x_1, which is a basic variable. The model contains the necessary information for this study, but it is much more detailed than the method used above.

9-11 SENSITIVITY ON PROCESSING RATES

To study the effects of changes in the hourly processing rates, it is necessary to examine the dual solutions corresponding to the constraints that define the hourly processing rates. The method is analogous to the study in Sec. 9.10. For example, if the rate of processing A through center 1 were increased from 300 to 301 gal/h, the dual solution indicates that there would be a dual solution of \$0.50, approximately. The constraint associated with this change is

$$x_1 - 300x_{13} = 0$$

Since the rate of processing is the coefficient of x_{13}, it is desirable to relate the change in rate of processing to a change in the right-hand side, which is evaluated by the dual solution. For this computation let the constraint be written

$$x_1 - 300x_{13} = b_1$$

Then

$$\Delta(x_1 - 300x_{13}) = \Delta b_1$$
$$\Delta x_1 - 300(\Delta x_{13}) = \Delta b_1$$

and since we wish to relate a change in b_1 to a change in x_{13} with x_1 held constant, we have

$$\Delta x_{13} = -\frac{\Delta b_1}{300}$$

Thus a change in x_{13} (hours of processing) is approximated by a change in b_9 divided by 300. As b_9 increases, the hours are decreased, which is equivalent to increasing the rate of processing.

Therefore, to approximate the effect of a change in the hourly rate, multiply the dual variable times the hours (solution for x_{13}). For this case the result is:

$$\text{approximate reduction} = 9.7466(0.50) = \$4.8733$$

In order to find the exact effect of a change from 300 to 301 in the rate of processing, we find that

$$x_{13} = \frac{x_1}{300} - \frac{x_1}{301}$$
$$= x_1\left(\frac{1}{300} - \frac{1}{301}\right)$$
$$= \frac{x_1}{(300)(301)}$$

When converted to dollars savings, it is multiplied by \$150, to obtain:

$$\text{savings} = \frac{x_1}{300}\frac{150}{301} = x_{13}(0.49834)$$

$$= \$4.8571$$

Thus the error is less than 2 cents on \$4.87.

In a similar fashion, all other rate-of-processing changes can be assessed by using the dual solutions corresponding to the constraint used for the definition. It is only necessary to multiply the dual solution by the corresponding hours in order to approximate the savings for an increase of 1 gal/h in processing rate. The errors are small.

9-12 SENSITIVITY ON PROCESSING COSTS

Another part of the analysis is to assess the effects of changing the processing costs for each of the centers. This can be done directly by a study of the ranging of the objective function for variables

$$x_{13} \text{ through } x_{20}$$

which are actually in the solution. The only one of these which is not in the solution is variable x_{17}. The effects of changing its coefficient must be obtained by evaluating its opportunity cost.

To evaluate the effects of changes in raw-material costs, the coefficients of variables x_1 and x_9 are analyzed, and the effects of changes in selling price are evaluated by looking at the coefficients of variables x_7, x_8, and x_{12}.

It should be evident that the larger and more detailed model provides much more managerial information.

9-13 EFFECTS ON THE COMPUTER

The smaller model resulted in three variables and seven constraints and, for the program used at Northwestern University's Computer Center, required 0.01 s of computing time. The larger model resulted in 44 variables and 24 constraints and required 0.1940 s of computing time. The computations were done on a CDC 6600.

The program used had good capabilities for sensitivity analysis. When use was made of ranging both the right-hand side and the coefficients of the objective function, the total computing times were 0.652 s and 2.673 s, respectively.

Memory requirements were obviously greater for the larger model, but they were no threat to the solution of the problem.

Exhibit 9-6

RNGRHS
******
(OPTIMALITY RANGE FOR RIGHT-HAND-SIDE CONSTANTS)
NON-SLACK RESOURCES ONLY

BI	XOUT	MIN BI Z-LOWER	ORIGINAL BI Z	MAX BI Z-UPPER	XOUT
1	38	-982.54 12335.	0. 18340.	2631.6 34421.	1
2	39	-358.91 15863.	0. 18340.	2500.0 35591.	1
3	39	-305.07 15605.	0. 18340.	2125.0 37391.	1
4	39	-127.88 16819.	0. 18340.	1700.0 38560.	4
5	8	0. 18340.	0. 18340.	1700.0 38560.	4
6	38	-1637.6 6330.8	0. 18340.	1176.5 26967.	9
7	38	-1391.9 5477.9	0. 18340.	1000.0 27580.	10
8	38	-1113.5 4642.7	0. 18340.	800.00 28180.	10
9	38	-1091.7 17794.	0. 18340.	2924.0 19802.	13
10	39	-377.80 18172.	0. 18340.	2631.6 19509.	14
11	41	-875.00 17710.	0. 18340.	2500.0 20140.	15
12	39	-335.82 18155.	0. 18340.	2125.0 19508.	16
13	8	0. 18340.	0. 18340.	2266.7 18238.	7
14	38	-1819.5 17248.	0. 18340.	1307.2 19124.	18
15	40	-4583.5 15952.	0. 18340.	1176.5 18952.	19
16	41	-1400.0 17500.	0. 18340.	1000.0 18940.	20
17	1	-2125.0 37391.	0. 18340.	305.07 15605.	39
22	43	1000.00 16655.	1700.0 18340.	1827.9 18647.	39
24	10	1700.0 13780.	2500.0 18340.	3200.0 22329.	43

CALCULATION TIME WAS .120 SECONDS

Exhibit 9-7

RNGOBO
******
(OPTIMALITY RANGE FOR COST COEFFICIENTS)
BASIC VARIABLES ONLY

CJ	XIN	MIN CJ ------- Z-LOWER	ORIGINAL CJ ----------- Z	MAX CJ ------- Z-UPPER	XIN
2	42	-1.5543	0.	9.8063	17
		14249.	18340.	44146.	
3	42	-1.6361	0.	10.322	17
		14249.	18340.	44146.	
4	42	-1.9248	0.	12.144	17
		14249.	18340.	44146.	
7	17	18.988	20.000	*INF*	
		16619.	18340.		
8		*INF*	20.000	21.012	17
			18340.	18340.	
10	44	-3.8758	0.	1.6361	42
		13780.	18340.	20264.	
11	44	-4.5598	0.	1.9248	42
		13780.	18340.	20264.	
12	44	12.300	18.000	20.406	42
		13780.	18340.	20264.	
1	42	-6.3989	-5.0000	3.8256	17
		14249.	18340.	44146.	
13	42	-569.66	-150.00	2497.7	17
		14249.	18340.	44146.	
14	42	-899.43	-200.00	4212.8	17
		14249.	18340.	44146.	
5	17	-.80960	0.	*INF*	
		16619.	18340.		
6		*INF*	0.	.75900	17
			18340.	18340.	
9	44	-9.4882	-6.0000	-4.5275	42
		13780.	18340.	20264.	
18	44	-2044.1	-300.00	436.24	42
		13780.	18340.	20264.	
19	44	-2110.4	-250.00	535.32	42
		13780.	18340.	20264.	
15	42	-589.02	-180.00	2400.6	17
		14249.	18340.	44146.	
16	17	-543.84	-220.00	*INF*	
		16619.	18340.		
20	44	-2063.9	-240.00	529.92	42
		13780.	18340.	20264.	

CALCULATION TIME WAS .109 SECONDS

The increases in both storage requirements and computing time were not insignificant but, on the other hand, they do not impose difficulties which cannot be overcome. Should the modeling of a problem along the lines illustrated in this example lead to excessive demands on either memory or computing time, the problem can be partitioned so as to relieve the excessive demands without diminishing the advantages discussed in the analysis, although more than one computer run might have to be made. One form of partitioning is shown below.

In this example, satisfactory reductions can be made by eliminating the definitions and constraints using variables x_{13} through x_{20}, and rewriting the center capacity constraints in terms of the remaining variables. These would then be

$$\begin{aligned} 5x_1 + 3x_9 &\leqq 24{,}000 \\ 8x_2 + 9x_5 &\leqq 43{,}200 \\ 48x_6 + 35x_{10} &\leqq 201{,}600 \\ 8x_3 + 5x_{11} &\leqq 32{,}000 \end{aligned}$$

and the objective function becomes

$$-5.5x_1 - 0.4444x_2 - 0.72x_3 - 0.55x_5 - 0.7143x_6 \\ +20x_7 + 20x_8 - 6.6x_9 - 0.5208x_{10} - 0.6x_{11} + 18x_{12}$$

This reduces the problem from one with 44 variables and 24 constraints to one with 28 variables and 16 constraints. Something is lost in the ability to make analyses in comparison with the model just discussed, but it does have ability to analyze recovery rates and sales income which the original small model did not have. It also provides for the analysis of all hourly processing rates except those associated with center 1, which were confounded with the two raw-material costs.

One can, by similar means, eliminate other portions of the problem and thus concentrate on one or two sections at a time. The summary computer output for each of the portions discussed for the large model are given here. Exhibits 9-4 and 9-5 were given along with the discussion in Sec. 9-10 and the *ranging* of the right-hand side and the *ranging* of the objective function are given in Exhibits 9-6 and 9-7.

9-14 CRITERIA FOR MODELING

In studying the previous example one might get the idea that it is necessary to increase both the number of constraints and the number of variables to get a more meaningful model. The material that follows is given to show that such is not the case. There appears to be no algorithm that can be applied to create the most meaningful model. To create an efficient model, one must have a good insight into the important facets of the problem.

Let us return now to the production scheduling problem which was solved in Chap. 8 with 27 variables and 26 constraints, and create a new model using an entirely different approach.

9-15 ALTERNATIVE FORMULATION WITH A REDUCED MODEL

The scheduling problem of Chap. 8 called for the production of products *A*, *B*, and *C* with the assignment of production of each of four stages to both day and night. We need to know how many of each product will be produced, so we let

x_1 = total number of product *A* to be produced
x_2 = total number of product *B* to be produced
x_3 = total number of product *C* to be produced

Now, realizing that one needs only to know how many hours to work on each stage, day and night, let

x_4 = number of hours of stage 1, day, to be used
x_5 = number of hours of stage 2, day, to be used
x_6 = number of hours of stage 3, day, to be used
x_7 = number of hours of stage 4, day, to be used
x_8 = number of hours of stage 1, night, to be used
x_9 = number of hours of stage 2, night, to be used
x_{10} = number of hours of stage 3, night, to be used
x_{11} = number of hours of stage 4, night, to be used

As before, there will be six constraints on quantities of *A*, *B*, and *C*. With simple variables representing the number of hours used for each of the stages, we will have simple constraints showing capacities. These will be $x_4 \leqq 800$, $x_5 \leqq 5000$, and so on. Four key constraints show the relation between the hours needed for production and the hours actually used, for each stage.

$$0.024x_1 + 0.04x_2 + 0.02x_3$$

is the number of hours used for the production of *A*, *B*, and *C* in stage 1, but this is also represented by $x_4 + x_8$. Therefore,

$$0.024x_1 + 0.04x_2 + 0.02x_3 - x_4 - x_8 = 0$$

There are similar constraints for the other three stages. The profit to be maximized is simply income less cost and is given by

$$0.75x_1 + 0.15x_2 + 0.10x_3$$

less the cost given by

$$0.1x_4 + 0.5x_5 + 0.3x_6 + 0.9x_7 + 0.3x_8 + 0.6x_9 + 0.5x_{10} + 0.9x_{11}$$

The mathematical model is therefore given by the relatively simple model:

Find

$$x_1, \ldots, x_{11} \geqq 0$$

such that

$$\begin{aligned}
x_1 &\geqq 10{,}000 \\
x_2 &\geqq 20{,}000 \\
x_3 &\geqq 10{,}000 \\
x_1 &\leqq 30{,}000 \\
x_2 &\leqq 120{,}000 \\
x_3 &\leqq 150{,}000 \\
x_4 &\leqq 800 \\
x_5 &\leqq 5000 \\
x_6 &\leqq 1{,}500 \\
x_7 &\leqq 600 \\
x_8 &\leqq 800 \\
x_9 &\leqq 5{,}000 \\
x_{10} &\leqq 1{,}500 \\
x_{11} &\leqq 600
\end{aligned}$$

$$\begin{aligned}
0.024x_1 + 0.04x_2 + 0.02x_3 - x_4 - x_8 &= 0 \\
0.6x_1 + 0.07x_2 + 0.04x_3 - x_5 - x_9 &= 0 \\
0.18x_1 + 0.01x_2 + 0.015x_3 - x_6 - x_{10} &= 0 \\
0.009x_1 + 0.03x_2 + 0.025x_3 - x_7 - x_{11} &= 0
\end{aligned}$$

and such that

$$z = 0.75x_1 + 0.15x_2 + 0.1x_3 - 0.1x_4 - 0.5x_5 - 0.3x_6 - 0.9x_7 - 0.3x_8 - 0.6x_9 - 0.5x_{10} - 0.9x_{11}$$

is a *maximum*.

This model has 11 variables and 18 constraints, in contrast with the initial formulation using 27 variables and 26 constraints. Do we get as much information from this solution? The actual computer input is given in Exhibit 9-8.

Exhibit 9-8

```
***** PROBLEM NUMBER  1 *****

        *L-TRONIC SMALL MODEL
        VARIABLES
        X1 TO X11
        MAXIMIZE
        .75X1+.15X2+.1X3-.1X4-.5X5-.3X6-.9X7-.3X8-.6X9-.5X10-.9X11
        CONSTRAINTS
    1.  X1.GE.10000
    2.  X2.GE.20000
    3.  X3.GE.10000
    4.  X1.LE.30000
    5.  X2.LE.120000
    6.  X3.LE.150000
    7.  X4.LE.800
    8.  X5.LE.5000
    9.  X6.LE.1500
   10.  X7.LE.600
   11.  X8.LE.800
   12.  X9.LE.5000
   13.  X10.LE.1500
   14.  X11.LE.600
   15.  .024X1+.04X2+.02X3-X4-X8=0
   16.  .6X1+.07X2+.04X3-X5-X9=0
   17.  .18X1+.01X2+.015X3-X6-X10=0
   18.  .009X1+.03X2+.025X3-X7-X11=0
        PRINT
        RNGOBJ
        RNGRHS
        OPTIMIZE
```

9-16 SOLUTION INFORMATION FROM THE REDUCED MODEL

Exhibit 9-9 gives the summary of results for this model. Interestingly enough, there are no alternate optimal solutions, since the only zeros in the opportunity cost column corresponding to nonbasic variables are for the artificial variables. What happened to the more than 18,000,000 alternative solutions developed in the model of Chap. 8?

The alternates are still there, but they are incorporated in the definition of the variables. Note that all of the 800 h available for stage 1 in the day are used. Therefore, any product may be assigned to these hours.

Exhibit 9-9

SUMMARY OF RESULTS

VAR NO	VAR NAME	ROW NO	STATUS	ACTIVITY LEVEL	OPPORTUNITY COST	BOUND VALUE
1	X1	--	B	12800.0000000	--	INF
2	X2	--	B	26560.0000000	--	INF
3	X3	--	B	11520.0000000	--	INF
4	X4	--	B	800.0000000	--	INF
5	X5	--	B	5000.0000000	--	INF
6	X6	--	B	1500.0000000	--	INF
7	X7	--	B	600.0000000	--	INF
8	X8	--	B	800.0000000	--	INF
9	X9	--	B	5000.0000000	--	INF
10	X10	--	B	1242.4000000	--	INF
11	X11	--	B	600.0000000	--	INF
12	--SLACK	1	B	2800.0000000	--	INF
13	--SLACK	2	B	6560.0000000	--	INF
14	--SLACK	3	B	1520.0000000	--	INF
15	--ARTIF D-	1	NB	--	0.0000000	INF
16	--ARTIF D-	2	NB	--	0.0000000	INF
17	--ARTIF D-	3	NB	--	0.0000000	INF
18	--SLACK D-	4	B	17200.0000000	--	INF
19	--SLACK D-	5	B	93440.0000000	--	INF
20	--SLACK D-	6	B	138480.0000000	--	INF
21	--SLACK D-	7	NB	--	.5803333	INF
22	--SLACK D-	8	NB	--	.5506667	INF
23	--SLACK D-	9	NB	--	.2000000	INF
24	--SLACK D-	10	NB	--	.5746667	INF
25	--SLACK D-	11	NB	--	.3803333	INF
26	--SLACK D-	12	NB	--	.4506667	INF
27	--SLACK D-	13	B	257.6000000	--	INF
28	--SLACK D-	14	NB	--	.5746667	INF
29	--ARTIF D-	15	NB	--	.6803333	INF
30	--ARTIF D-	16	NB	--	1.0506667	INF
31	--ARTIF D-	17	NB	--	.5000000	INF
32	--ARTIF D-	18	NB	--	1.4746667	INF

MAXIMUM VALUE OF THE OBJECTIVE FUNCTION = 6764.800000

CALCULATION TIME WAS .0850 SECONDS FOR 14 ITERATIONS.

Similarly, all 800 h for stage 1 in the night are used again with the freedom of assigning *A*, *B*, and *C* within the overall constraints of the problem. With this solution, then, the foreman in charge of assigning production may start with any one of the products on stage 1 of the day shift and continue any set of assignments of *A*, *B*, and *C* as long as the total production of *A* is equal to 12,800, that of *B* is equal to 26,560, and that of *C* is equal to 11,250. Thus although the mathematical model does not lead to any alternate solutions, the very definition of the variables themselves implies alternates.

The dual solutions appear for each of the meaningful constraints and therefore nothing is lost. Furthermore, the sensitivity on the coefficients of the objective function is much more direct, since the analysis is on the actual income and cost parameters themselves. The computer output for these analyses is shown in Exhibits 9-10 and 9-11.

Exhibit 9-10

RNGRHS
******
(OPTIMALITY RANGE FOR RIGHT-HAND-SIDE CONSTANTS)
NON-SLACK RESOURCES ONLY

BI	XOUT	MIN BI ------- Z-LOWER	ORIGINAL BI ----------- Z	MAX BI ------- Z-UPPER	XOUT
7	13	699.18	800.00	819.69	14
		6706.3	6764.8	6776.2	
8	14	4050.0	5000.0	5791.8	27
		6241.7	6764.8	7200.8	
9	27	1242.4	1500.0	2742.4	10
		6713.3	6764.8	7013.3	
10	14	584.68	600.00	733.70	13
		6756.0	6764.8	6841.6	
11	13	699.18	800.00	819.69	14
		6726.5	6764.8	6772.3	
12	14	4050.0	5000.0	5791.8	27
		6336.7	6764.8	7121.6	
14	14	584.68	600.00	733.70	13
		6756.0	6764.8	6841.6	
15	13	-100.82	0.	19.689	14
		6696.2	6764.8	6778.2	
16	14	-950.00	0.	791.80	27
		5766.7	6764.8	7596.7	
17	27	-257.60	0.	1242.4	10
		6636.0	6764.8	7386.0	
18	14	-15.323	0.	133.70	13
		6742.2	6764.8	6962.0	

CALCULATION TIME WAS .061 SECONDS

Exhibit 9-11

RNGOBJ
******
(OPTIMALITY RANGE FOR COST COEFFICIENTS)
BASIC VARIABLES ONLY

CJ	XIN	MIN CJ Z-LOWER	ORIGINAL CJ Z	MAX CJ Z-UPPER	XIN
10	24	-1.1863 5912.1	-.50000 6764.8	-.30000 7013.3	23
11	28	-1.4747 6420.0	-.90000 6764.8	*INF*	
3	24	.94207E-01 6698.1	.10000 6764.8	.10493 6821.6	25
2	25	.14415 6609.5	.15000 6764.8	.16171 7075.9	24
7	24	-1.4747 6420.0	-.90000 6764.8	*INF*	
9	26	-1.0507 4511.5	-.60000 6764.8	*INF*	
8	25	-.68033 6460.5	-.30000 6764.8	*INF*	
1	26	.49650 3520.0	.75000 6764.8	.90559 8756.4	25
4	21	-.68033 6300.5	-.10000 6764.8	*INF*	
5	22	-1.0507 4011.5	-.50000 6764.8	*INF*	
6	23	-.50000 6464.8	-.30000 6764.8	*INF*	

CALCULATION TIME WAS .055 SECONDS

9-17 EFFICIENCY OF LINEAR PROGRAMMING MODELS

In Chaps. 8 and 9, illustrations were given for improving the modeling for a linear programming problem. In the case of the scheduling problem, requiring processing through several centers, the natural instinct is to use as few variables as possible and restrict yourself to the seven natural constraints of the problem. When this is done, you of course get the solution to the scheduling problem, but it is not particularly enlightening in terms of sensitivity analysis. In fact, it revealed nothing about recovery rates or processing rates, nor was the objective function in a form that revealed anything about costs on an individual basis.

When the problem was expanded in both the number of constraints and the number of variables, much greater insight was gained into the problem.

Influences of changes in the recovery rates and the processing rates could be isolated and analyzed easily. The objective function is in a form in which one can analyze the effect of changes in costs of processing rates, costs of raw materials, and the sales price of each of the final products.

Thus we found that considerable expansion of the problem was a distinct advantage, at very little cost in computing time. The better model was not really the natural one.

In the scheduling problem requiring the manufacturing to go through four stages of production for each of the three products, the better model was the one with fewer variables and fewer constraints. The initial model, which was given in Chap. 8, detailed the complete production schedule by product, by stage, and by shift. This results in a large number of alternates and thus to investigate the desirability of any particular solution you must enter the appropriate variables into the program, which is not a particularly simple thing to do. Many programs do not allow for this capability, and thus it must be done manually. If the foreman on the job were handed this basic solution and told that there were alternates but that he or she would have to find them, he would probably just use the initial solution given to him.

The second model for this case had fewer variables and fewer constraints, and, strangely enough, no alternates. The alternates are embedded within the definition of the variables. The solution does not tell which product is assigned to which stage on which shift; it only tells the hours used on each stage. Thus when the foreman is presented with a schedule, he or she has a number of alternates, but they are very easily specified. He can essentially put anything he wants on stage 1 in the day as long as he stops assigning when he has used up the 800 h, and as long as he stops when he reaches the exact amount for any one of the products, or if he has not reached the proper production level when he runs out of hours, he produces the remainder on stage 1 at night. Thus he has great flexibility, and the results are given in language which he can understand and for which he can make the necessary computations very easily. Furthermore, the sensitivity analysis is direct and meaningful for this form of the model.

From these two illustrations one can see the importance of "proper" modeling. One cannot necessarily specify what is meant by "proper." In the one case it is an expanded model, and in the other case it is a reduced model, so size itself is not a criterion. In fact, it probably is somewhat of an art. There are differences in what you can get from any specified model, so care should be given to its formulation.

9-18 ADDITIONAL MODEL CHARACTERISTICS

The material that has been discussed up to this point should enable one to solve all straightforward linear programming problems. There are even some characteristics of apparently simple models that escape the attention of some of the more sophisticated linear programming experts. For example, one should not take as gospel the material on ranging of the right-hand side in the case in which there are degenerate solutions. The reason for this is that with some other variable at the zero level, it may result in a different range. It is possible to determine which basic solution will give the maximum range, but I know of no program that will do this automatically. It can be said, however, that the range is at least as large as that given by the program, so that if one determines several ranges, he or she can be sure that the range is at least as large as the largest of those obtained. A routine can be established for doing this, but it has not been done here. Suffice it to say that one should be wary of ranging results in cases of degeneracies. This is really not just an idle threat, since real-world problems very often have degeneracies.

One often finds that real-world problems have an **A** matrix with many zeros—we call it a sparse matrix. In some of these cases it is possible to dissect the problem into several smaller linear programming problems, all of which are connected by one or more constraints and an overlapping objective function. As an illustration, one might be interested in the optimal operation of a multiplant company. The individual linear programming problems might be the optimization of each of the plants. The connecting constraints would be the home office controls relating to multiplant operation. When these are applied, the individual plant optimizations may no longer be feasible and therefore may require another iteration. The method of handling this kind of a solution is known as the *decomposition principle.* The revised simplex method is used in this solution but is not discussed here. The method of solution is not only faster than solving the problem as one large problem, but it may also enable one to fit a problem within computer memory when it would not be possible for a large model.

I think it is safe to say that there is literally no end to the ramifications and ingenious solutions and models available for linear programming problems. The material given in this text should give readers a substantial background for solving linear programming problems. Only experience can improve one's sophistication in handling such problems.

9-19 GOAL PROGRAMMING

During the process of solving linear programming problems, it becomes evident that there is often some kind of a conflict that causes difficulty in setting up the problem. The normal linear programming problem will have a series of constraints, some of which are binding and some of which are not. It usually is not known which will be binding when the linear programming model is formulated, for if it were, the nonbinding constraints would not be in the formulation unless they were desired for purposes of sensitivity analysis.

An inexperienced person setting up a linear programming model might well think that it would be necessary to make sure that the set of constraints is consistent in order to pursue the simplex method. Of course, a problem has no solution if the constraints are inconsistent, but this can be discovered in the processing of trying to obtain a solution. If the results show that there is no "feasible solution" space, i.e., no solution, then it is possible either to change the constraints so that they are consistent or to use the concept of "goal programming" if it does not appear desirable to relax any of the constraints or goals.

In addition to the fact that there may be no feasible solution, it may not be clear that the objective function is unique. Firms often seek profit maximization while pursuing other noneconomic goals. Linear programming does not have the ability to deal with this complex reality. To overcome the one-dimensional nature of the objective function in linear programming, efforts are made to convert various goals, costs, or value measurements into one criterion, namely, utility. Linear programming basically requires maximization of profit, minimization of costs, or some other objective criterion. Even when decision makers utilize linear programming, implicitly they must weigh the output of the decision against some other objectives they could not include in the model. This could lead to a much more complex programming model. Goal programming is one way to handle some of the difficulties. It is therefore a modification and extension of linear programming. The method allows a simultaneous solution of a system of complex objectives, even though it be composed of nonhomogeneous units of measure. The solution of the problem requires the establishment of an order of importance among these competing goals, so that the lesser important goals are considered only after the more important goals are either satisfied or have reached a point beyond which no further improvement is deemed necessary.

The goal-programming objective function usually contains variables which are either + or − deviations from the established goal or subgoal. For illustration, let us consider the TV production problem from Chap. 1. The mathematical statement of that problem was

Find

$$x_1, x_2 \geqq 0$$

such that

$$x_2 \leqq 50$$

$$6x_1 + 18x_2 \leqq 1800$$

$$x_1 + 1.6x_2 \leqq 240$$

$$0.5x_1 + 2x_2 \leqq 162$$

and such that

$$z = 15x_1 + 45x_2 \text{ is a } maximum$$

In this problem, the first constraint is on color-tube production, the second is on hours in the chassis assembly department, the third is on hours in the set assembly department, and the fourth is on hours in the inspection department.

Let us now consider a modification to the problem in view of the multiple objectives. In order of priority these modifications are as follows:

1. First priority is to obtain a profit as close to \$4300 as possible.
2. The second priority is to produce as close to 50 color sets or color tubes as possible.
3. The third priority is to come as close as possible to using all of the 1800 h in the chassis assembly department.

With these goals in mind, let us consider the new variables

d_1^- = the amount by which the profit falls short of the goal of \$4300.
d_1^+ = the amount by which the profit exceeds the goal of \$4300.
d_2^- = the amount by which the color-set production falls short of the goal of 50.
d_3^- = the amount by which the chassis assembly hours fall short of the 1800 hr available.
P_1, P_2, P_3 = priority weights assigned to each of the deviations with the condition that $P_1 > P_2 > P_3$.

Our goal-programming problem statement then becomes
Find

$$x_1, x_2, d_1^-, d_1^+, d_2^-, d_3^- \geqq 0$$

such that

$$\begin{aligned} 15x_1 + 45x_2 + d_1^- - d_1^+ &= 4300 \\ x_2 + d_2^- &= 50 \\ 6x_1 + 18x_2 + d_3^- &= 1800 \\ x_1 + 1.6x_2 &\leqq 240 \\ 0.5x_1 + 2x_2 &\leqq 162 \end{aligned}$$

and such that

$$Z = P_1 d_1^- + P_1 d_1^+ + P_2 d_2^- + P_3 d_3^- \text{ is a } minimum$$

To solve this problem, the techniques of the simplex method are modified. The coefficients of the objective function are stated in terms of the priorities P_i, which are treated in somewhat the same fashion as the big M in the case of the artificial variable.

If we set up the ordinary simplex tableau, we have

Basis	C_j	0	0	P_1	P_1	P_2	P_3	0	0	
		x_1	x_2	d_1^-	d_1^+	d_2^-	d_3^-	s_1	s_2	b
d_1^-	P_1	15	45	1	−1	0	0	0	0	4300
d_2^-	P_2	0	1	0	0	1	0	0	0	50
d_3^-	P_3	6	18	0	0	0	1	0	0	1800
s_1	0	1	1.6	0	0	0	0	1	0	240
s_2	0	0.5	2	0	0	0	0	0	1	162

The first feasible solution is

$$d_1^- = 4300$$

$$d_2^- = 50$$

$$d_3^- = 1800$$

$$s_1 = 240$$

$$s_2 = 162$$

with

$$Z = 4300P_1 + 50P_2 + 1800P_3$$

Since there are only three nonbasic variables, we need compute only three $(z_j - c_j)$'s. These are

$$z_1 - c_1 = 15P_1 + 6P_3$$

$$z_2 - c_2 = 45P_1 + P_2 + 18P_3$$

$$z_4 - c_4 = -2P_1$$

Since we are minimizing the objective function, we wish to select the largest of these $(z_j - c_j)$'s. Since P_1 represents the highest priority, $z_2 - c_2$ is the largest of these and, therefore, x_2 is chosen for entry into the basis. The pivot point is chosen in the regular way so that x_2 will replace d_2^-. After this step is completed, we have

$$z_1 - c_1 = 15P_1 + 6P_3$$

$$z_4 - c_4 = -2P_1$$

$$z_5 - c_5 = -45P_1 - P_2 - 18P_3$$

from which we see that x_1 should be brought into the solution, replacing s_2. Continuing this process, we shall select the variable for entry into the basis by finding the largest $z_j - c_j$. Since $P_1 > P_2 > P_3$, we shall continue to get improvement as long as any $z_j - c_j$ contains a positive coefficient for any P_1 term. When all such positive terms have been eliminated, we have

	C_j	0	0	P_1	P_1	P_2	P_3	0	0	
Basis		x_1	x_2	d_1^-	d_1^+	d_2^-	d_3^-	s_1	s_2	b
d_2^-	P_2	0	0	$\frac{1}{15}$	$-\frac{1}{15}$	1	0	0	-2	$12\frac{2}{3}$
x_2	0	0	1	$-\frac{1}{15}$	$\frac{1}{15}$	0	0	0	2	$37\frac{1}{3}$
d_3^-	P_3	0	0	$-\frac{2}{5}$	$\frac{2}{5}$	0	1	0	0	80
s_1	0	0	0	$-\frac{4}{25}$	$\frac{4}{25}$	0	0	1	2.8	$5\frac{3}{5}$
x_1	0	1	0	$\frac{2}{5}$	$-\frac{2}{5}$	0	0	0	-6	$174\frac{2}{3}$

with

$$z_3 - c_3 = -P_1 + \tfrac{1}{15}P_2 - \tfrac{2}{5}P_3$$

$$z_4 - c_4 = -P_1 - \tfrac{1}{15}P_2 + \tfrac{2}{5}P_3$$

$$z_8 - c_8 = -2P_2$$

Since $z_3 - c_3$ has the positive term $P_2/15$, we could enter d_1^-, which means that we would not reach the highest priority goal of \$4300 profit. Since $z_4 - c_4$ has the positive term $\frac{2}{5}P_3$, we could enter d_1^+ which means that we would overshoot the profit of \$4300. Therefore if we are to meet the highest priority of \$4300 profit, our solution is

$$x_1 = 174\tfrac{2}{3}$$

$$x_2 = 37\tfrac{1}{3}$$

$$\text{Profit} = \$4300$$

$$d_2^- = 12\tfrac{2}{3}$$

$$d_3^- = 80$$

The foregoing illustration has been given to present another way in which to model a problem.

EXERCISES

No exercises are given here. We suggest that you review some of the more difficult problems given throughout the text and see if you can construct more sophisticated models, since you now have a much better appreciation of what is involved in the solution of linear programming problems.

BIBLIOGRAPHY

This bibliography is strictly for reference purposes and is not intended to be a complete list of either theory or applications in the area of linear programming. A number of standard references may give the reader a different viewpoint on a topic of particular interest.

1. Beale, E. M. L.: *Mathematical Programming in Practice*, John Wiley & Sons, Inc., New York, 1968.
2. Charnes, A., and W. W. Cooper: *Management Models and Industrial Applications of Linear Programming*, 2 vols., John Wiley & Sons, Inc., New York, 1960.
3. Chung, A.: *Linear Programming*, Charles E. Merrill Publishing Company, Columbus, Ohio, 1963.
4. Dantzig, G. B.: *Linear Programming and Extensions*, Princeton University Press, Princeton, N.J., 1963.
5. Dorfman, R., P. A. Samuelson, and R. M. Solow: *Linear Programming and Economic Analysis*, McGraw-Hill Book Company, New York, 1958.
6. Driebeek, N. J.: *Applied Linear Programming*, Addison-Wesley Publishing Company, Inc., Reading, Mass., 1969.
7. Ford, L. R., Jr., and D. R. Fulkerson: *Flows in Networks*, Princeton University Press, Princeton, N.J., 1962.
8. Garvin, W. W.: *Introduction to Linear Programming*, McGraw-Hill Book Company, New York, 1960.

†9. Gass, S. I.: *Linear Programming: Methods and Applications*, 4th ed., McGraw-Hill Book Company, New York, 1975.

10. Hadley, G. F.: *Linear Programming*, Addison-Wesley Publishing Company, Inc., Reading, Mass., 1962.
11. Heady, E. O., and W. Candler: *Linear Programming Methods*, Iowa State College Press, Ames, Iowa, 1958.

†12. Kim, C.: *Introduction to Linear Programming*, Holt, Rinehart and Winston, Inc., New York, 1971.

† These two books have rather comprehensive bibliographies. In each case the texts are divided into various categories. In addition to the book references, there is a comprehensive bibliography of articles on both theory and application. The bibliographies are rather recent and include everything that is good up to perhaps two years ago.

13. Lasdon, Leon: *Optimization Theory for Large Systems*, Macmillan Publishing Co., Inc., New York, 1970.
14. Loomba, N. P.: *Linear Programming, An Introductory Analysis*, McGraw-Hill Book Company, New York, 1964.
15. Luenberger, D. G.: *Introduction to Linear and Nonlinear Programming*, Addison-Wesley Publishing Company, Inc., Reading, Mass., 1973.
16. Orchard-Hays, W.: *Advanced Linear-Programming Computing Techniques*, McGraw-Hill Book Company, New York, 1968.
17. Simonnard, M.: *Linear Programming*, Prentice-Hall, Inc., Englewood Cliffs, N.J., 1966.
18. Smythe, W. R., Jr., and L. A. Johnson: *Introduction to Linear Programming*, Prentice-Hall, Inc., Englewood Cliffs, N.J., 1966.
19. Sweeney, D. J., D. R. Anderson, and T. A. Williams: *Linear Programming for Decision Making*, West Publishing Company, New York, 1974.
20. Vadja, S.: *Readings in Linear Programming*, John Wiley & Sons, Inc., New York, 1958.
21. Vadja, S.: *Readings in Mathematical Programming*, John Wiley & Sons, Inc., New York, 1962.
22. Vadja, S.: *The Theory of Games and Linear Programming*, Methuen & Company Ltd., London, 1956.

INDEX

Array, 19
Assignment problems, 158

Basis set of vectors, 40–42
  inverse of basis set, 79–80
  replacement of vector with an arbitrary vector, 42–43

Chemical process problem, 180–200
  advantages of second model, 191
  alternative statement and model, 188–191
  computer costs, 197
  detailed sensitivity, 192–197
  efficiency of model, 186–187
  intuitive model, 182–183
  mathematical model, 183–185
  optimal solution, 185–186
  scheduling parameters, 190–191

Computer:
  MPOS program, 165–174
    (*See also* MPOS program)
  scheduling example, 161–173
    mathematical model, 162–165
  use in solving linear programming problems, 161
Constraints, 1–3
  equal-to, 83, 84
  equal-to-or-greater-than, 17, 27, 83, 84
  equal-to-or-less than, 17, 58
Convex set, 52

Dantzig, G. B., 1
Dual linear programming problem, 16, 17
  computation of dual solution from primal solution, 107–108
  in matrix notation, 30, 31, 103

Dual linear programming problem:
method of writing dual statement from primal statement, 110–112
properties of, 104–107
relation to primal, 17, 18
solutions: for ≦ constraints, 108–109
for ≧ constraints, 109
for = constraints, 109–110
summary of dual solutions, 110
Dual simplex method, 124–125

E-matrix, 134, 136, 138, 139
Euclidean space, 36
Extreme point, 52

Four-stage production problem:
alternative model, 201–203
efficiency of model, 206–207
intuitive model, 161–165

Generalized transportation model, 154–155
Goal programming, 208–212
Graphical solutions:
alternate optima, 5
maximization, 2
minimization, 6
unbounded feasible region, 8

Hyperplanes, 51–52

Interpretation of solution, 5
Inverse:
illustrative example of, 47–48

Inverse:
matrix, 31–32
product form of, 45–46

Linear function, 2
Linear independence, 36–38
comments on, 72–73
properties of, 39–40
Linear programming:
basic feasible solution, 60–62
comments on solution, 95–96
concept: of feasible solution, 53
of optimal basic feasible solution, 53–54
of solution, 53
definition, 1
generalized problems, 83–98
introductory problem, 13–15
in matrix notation, 25, 26, 59
minimization, 92–94
origin of, 1
primal problem, 15–16
simplex method, 58
summary of solution method, 80–81
unbounded solution, 95
Linear systems of equations:
basic feasible solution, 60–61
basic solutions, number of, 49, 50
degenerate solution, 51
feasible solution, 51
nonfeasible solution, 51
solution: by Gaussian elimination, 32–33
by inverse, 32, 49
of multiple systems, 34

Manufacturing mix problem, 74–76
Mathematical model, 2, 3, 5–7, 9, 10
criteria for, 200–201
for interpretation, 180–212
model characteristics, 207–208

Matrix(ces), 18
  addition, 19–21
  associative law: for addition, 19
    for multiplication, 25
  commutative law for addition, 19
  definition, 19
  distributive law, 25
  E-, 134–139
  equality of, 19
  identity, 28, 61
  multiplication of, 24–25
    by scalar, 20, 21
    by vector, 23
  null, 29
  partitioned, 131–132
  rank of: column, 44
    row, 44
  square, 19
  transposed, 29
Matrix inverse, 31–36
Matrix operations, examples of, 20
MPOS program, 165–174
  alternate optimal solutions, 169–170
  coefficients of objective function, 171–173
  input format, 165–166
  optimal solution—basis inverse, 173
  right-hand side, 170–171
  solution results, 166
    opportunity costs, 168–169
  summary of results, 167–168

n-dimensional space, 20
Northwest corner rule, 147, 148, 152, 155

Objective function:
  alternative logic, 79
  use of, as additional constraint, 76–78

Partitioned matrices, 131–132
Point sets, 51
  closed half-space, 52
  convex set, 52
  extreme point, 52
  open half-space, 52
Primal problem 15–16
Product mix problem, 14

Revised simplex procedure, 130–131
  condensed summary of tableaux, 141–142
  condensed tableaux, 135–141
  revised tableaux, 131
  selection of incoming variable, 132–133
  selection of outgoing variable, 133–134
  summary of method, 142–143
  updated basis inverse, 134

Second penalty method, 155–158
Sensitivity analysis, 102–127
  adding new constraint, 123–125
  addition of new variable, 120
  changes: in **A** matrix, 121–123
    in *c* vector, 118–120
    in right-hand side, 112–117
  computer instructions for, 170–173
  dual simplex method, 124–125
  duality, importance of, 102–104
  removal of constraint, 125–127
  summary, 127–128
Simplex method, 58–81
  computer solution, 72
  flow chart, 71
  general format of solution, 108
  initial tableau, 63
  iteration, 67
  optimal tableau, 69
  pivot column, 65

Simplex method:
pivot element, 65
pivot row, 65
selection of incoming vector, 64
selection of outgoing vector, 64
summary tableaux, 70
test for optimality, 69
Solution, 5
Solution space, 3
Steel fabricating problem, 96–99
Straight lines, 52

Transportation model, 144–145
alternate optima, 153–154
applications, 151
degeneracies, 152–153
dual statement, 145–147
dual variables, 146, 148–151
efficient first feasible, 155–158
flow chart, 152
generalized model, 154–155
initial tableau, 148
introduction, 144
northwest corner rule, 147–148
second penalty method, 155–158
solution routine, 147–152
tableau revision, 149–151
test for improvement, 149

Two-phase method, 89–92

Variables:
artificial, 49, 84–89
basic, 50
big *M*, 86
nonbasic, 50
slack, 49, 59, 83
surplus, 84
Vectors:
addition, 20, 21
basis set of, 40–41
column, 20
directed-line-segment approach, 21
inequality, 22
linear combination, 37, 62
linear dependence, 36, 38
linear independence, 36, 38
multiplication by scalar, 20, 21
nonbasic variable, 62–63
null, 22
number of, in basis, 44
row, 20
scalar product of two, 22–23
spaces and subspaces, 44
spanning set, 40–41
special properties of, 39–40
unit, 22